U.S. Department
of Transportation
**National Highway
Traffic Safety
Administration**

DOT HS 811 464　　　　　　　　　　　　　　　　　　　　　　May 2011

Integrated Vehicle-Based Safety Systems

Heavy-Truck Field Operational Test Independent Evaluation

This publication is distributed by the U.S. Department of Transportation, National Highway Traffic Safety Administration, in the interest of information exchange. The opinions, findings and conclusions expressed in this publication are those of the author(s) and not necessarily those of the Department of Transportation or the National Highway Traffic Safety Administration. The United States Government assumes no liability for its content or use thereof. If trade or manufacturers' names or products are mentioned, it is because they are considered essential to the object of the publication and should not be construed as an endorsement. The United States Government does not endorse products or manufacturers.

REPORT DOCUMENTATION PAGE			Form Approved OMB No. 0704-0188
	Public reporting burden for this collection of information is estimated to average 1 hour per response, including the time for reviewing instructions, searching existing data sources, gathering and maintaining the data needed, and completing and reviewing the collection of information. Send comments regarding this burden estimate or any other aspect of this collection of information, including suggestions for reducing this burden, to Washington Headquarters Services, Directorate for Information Operations and Reports, 1215 Jefferson Davis Highway, Suite 1204, Arlington, VA 22202-4302, and to the Office of Management and Budget, Paperwork Reduction Project (0704-0188), Washington, DC 20503.		
1. AGENCY USE ONLY (Leave blank)	2. REPORT DATE May 2011		3. REPORT TYPE AND DATES COVERED Final Report August 2006-August 2010
colspan="3"	4. TITLE AND SUBTITLE Integrated Vehicle-Based Safety Systems Heavy-Truck Field Operational Test Independent Evaluation	5. FUNDING NUMBERS Inter-Agency Agreement HS-22A1 DTNH22-08-V-00015
	6. AUTHOR(S) Emily Nodine, Andy Lam, Wassim Najm, Bruce Wilson, and John Brewer	
colspan="3"	7. PERFORMING ORGANIZATION NAME(S) AND ADDRESS(ES) U.S. Department of Transportation Research and Innovative Technology Administration John A. Volpe National Transportation Systems Center Cambridge, MA 02142	8. PERFORMING ORGANIZATION REPORT NUMBER DOT-VNTSC-NHTSA-11-01
colspan="3"	9. SPONSORING/MONITORING AGENCY NAME(S) AND ADDRESS(ES) John J. Ference U.S. Department of Transportation National Highway Traffic Safety Administration	10. SPONSORING/MONITORING AGENCY REPORT NUMBER DOT HS 811 464
	11. SUPPLEMENTARY NOTES
colspan="3"	12a. DISTRIBUTION/AVAILABILITY STATEMENT This document is available to the public through the National Technical Information Service www.ntis.gov.	12b. DISTRIBUTION CODE
	13. ABSTRACT (Maximum 200 words) This report presents results from the independent evaluation of a field operational test using a fleet of heavy trucks outfitted with a prototype integrated crash warning system. This effort was conducted as part of the U.S. DOT's Integrated Vehicle-Based Safety Systems (IVBSS) program. The system tested included rear-end, lane-change/merge, and lane departure crash warning functions. The goals of the evaluation were to assess safety impact and driver acceptance, and to characterize the prototype system's warning capabilities. Eighteen volunteer drivers from a commercial fleet operated the 10 equipped heavy trucks, accumulating 600,000 miles over a 10-month period. The test period consisted of 2 months of baseline driving, when the system was disabled, and an 8-month treatment period, when the system was enabled and warnings were presented to the drivers. Comparisons were made between baseline driving and the treatment period to determine the effect of system use on driver behavior and performance. Results from driver debriefings and surveys indicated that they had a favorable impression of the prototype system they used during the field trial, reporting that it was easy to learn and use. A majority of the drivers also said they would prefer driving a truck with the integrated system over a conventional, unequipped truck. Aside from improvements in lane-keeping behavior and turn signal use, results from the field test indicate that between 3,000 and 13,000 target crashes could be prevented annually by full deployment of integrated safety systems in the U.S. heavy-truck fleet.
colspan="3"	14. SUBJECT TERMS Commercial heavy-truck safety, crash warning, driver-vehicle interface, intelligent vehicles, driving conflicts, driver acceptance, driving performance measures, Integrated Vehicle-Based Safety Systems (IVBSS), near-crashes, system capability	15. NUMBER OF PAGES 140
			16. PRICE CODE
17. SECURITY CLASSIFICATION OF REPORT Unclassified	18. SECURITY CLASSIFICATION OF THIS PAGE Unclassified	19. SECURITY CLASSIFICATION OF ABSTRACT Unclassified	20. LIMITATION OF ABSTRACT Unlimited

NSN 7540-01-280-5500

Standard Form 298 (Rev. 2-89)
Prescribed by ANSI Std. 239-18
298-102

Table of Contents

List of Figures v
List of Tables vii
List of Acronyms viii
Executive Summary 1

1. **Introduction** 6
   1.1 Integrated Safety System 6
   1.2 Target Crashes 7
   1.3 Field Operational Test 9
      1.3.1 Demographics of Field Test Participants 10
      1.3.2 Summary of Field Test Exposure 11
   1.4 Independent Evaluation 12
      1.4.1 Data Processing 13
      1.4.2 Multimedia Data Analysis Tool 14

2. **Safety Impact** 16
   2.1 Safety Impact Technical Approach 16
   2.2 Overall Driving Experience 17
      2.2.1 Speed Maintenance 18
      2.2.2 Headway Keeping 19
      2.2.3 Lane-Change Behavior 20
      2.2.4 Lane Keeping 21
      2.2.5 Attention to Primary Driving Task 22
   2.3 Driving Conflict Experience 28
      2.3.1 Exposure to Driving Conflicts 29
      2.3.2 Driver Response to Driving Conflicts 31
   2.4 Near-Crash Experiences 32
   2.5 Projection of Potential Safety Benefits 38

3. **Driver Acceptance** 41
   3.1 Driver Acceptance Technical Approach 41
      3.1.1 Acceptance by Driver and Objective 41
      3.1.2 Acceptance by Independent Variables 42
   3.2 Survey Results 45
      3.2.1 General Feedback 45
      3.2.2 Ease of Use 46
      3.2.3 Perceived Usefulness 50
      3.2.4 Ease of Learning 53
      3.2.5 Advocacy 54
      3.2.6 Driving Performance 57
   3.3 Driver Acceptance by Demographic Variables 58
   3.4 Driver Acceptance by Driver Experience Variables 60

4. **System Capability** 63
   4.1 Sensors 63
      4.1.1 Forward-Looking Sensors 63

4.1.2	Side-Looking Sensors	65
4.1.3	Lane Tracking	68
4.2	Warning Logic	70
4.2.1	Hazard Propensity	71
4.2.2	Driver Response	74
4.2.3	Comparison of Alert Rates Between Baseline and Treatment	78
4.3	Driver-Vehicle Interface	81
4.4	System Robustness	84

5. **Conclusions** 87
   References 90

Appendix A: Post-Drive Survey 91
Appendix B: Data Processing and Data Mining 105
Appendix C. Post-Drive Survey Mapping to Acceptance Objectives 110
Appendix D: Video Analysis 111
Appendix E. Video Coding Manual 114
Appendix F. Conflict Identification Thresholds 127

List of Figures

Figure 1.	Driver-vehicle interface of the integrated safety system	7
Figure 2.	Data processing procedures	13
Figure 3.	Screen view of multimedia data analysis tool	15
Figure 4.	Safety benefits framework	17
Figure 5.	Change in percent of alerts with secondary tasks from baseline to treatment	24
Figure 6.	Change in the average number of secondary tasks per alert from baseline to treatment	25
Figure 7.	Change in percent of alerts with eyes-off-forward-scene from treatment to baseline	27
Figure 8.	Change in number of near-crashes encounters per 1,000 miles in baseline versus treatment	35
Figure 9.	Distribution of near-crashes by type and threat validity	36
Figure 10.	Change in specific near-crash rates between baseline and treatment	36
Figure 11.	Aggregate results of 12 survey items related to ease of use	47
Figure 12.	Responses to the question "How satisfied were you with the integrated system?"	47
Figure 13.	Responses to the statement "The integrated system made my job easier"	48
Figure 14.	Drivers' opinions of the integrated system auditory warnings	49
Figure 15.	Drivers' responses to the survey item "The auditory alerts were not annoying"	49
Figure 16.	Aggregate results of 10 survey items related to perceived usefulness	50
Figure 17.	Drivers' opinions on the safety increase associated with the system	51
Figure 18.	Responses to "The number of false warnings caused me to ignore the integrated system"	52
Figure 19.	Responses to "The integrated system gave me alerts when I did not need them"	53
Figure 20.	Responses to the question "I always knew what to do when the system provided me with a warning"	54
Figure 21.	Drivers' opinions of having new technology in their trucks	55
Figure 22.	Drivers' willingness to drive with the integrated system	56
Figure 23.	Driver's willingness to endorse the integrated system to their employer	57
Figure 24.	Drivers' self-reported reliance on the integrated system	58
Figure 25.	Proportion of FCW alerts triggered by out-of-path targets	64
Figure 26.	Distribution of out-of-path target types in FCW-S alerts	64
Figure 27.	Frequency rates of FCW alerts triggered by out-of-path targets	65
Figure 28.	Proportion of LCM and LDW-I alerts triggered by targets not in adjacent lanes	66
Figure 29.	Distribution of non-adjacent target types for LCM and LDW-I alerts	67
Figure 30.	Frequency rates of LCM and LDW-I alerts triggered by non-adjacent targets	68
Figure 31.	Distribution of LDW-C alerts by occurrence of lane excursion	69
Figure 32.	Probability of LDW-C alerts without lane excursion under various conditions	70
Figure 33.	Mapping of valid alerts to driving conflicts	72
Figure 34.	Mapping of valid alerts to near-crashes	73
Figure 35.	Proportion of valid alerts associated with secondary tasks and eyes-off-forward-scene	73
Figure 36.	Responses to survey items about drivers' reaction to nuisance alerts	74
Figure 37.	Drivers' opinions of the system's issuance of false warnings by alert type	75
Figure 38.	Breakdown of driver action in response to valid alerts	76
Figure 39.	Average brake reaction time to FCW alerts between baseline and treatment	77

Figure 40. Average peak deceleration level to FCW alerts between baseline and treatment 77
Figure 41. Average peak lateral acceleration to side alerts between baseline and treatment....... 78
Figure 42. Percent change in FCW-M alert rates between baseline and
            treatment test condition.. 80
Figure 43. Percent change in LDW-C left alert rates between baseline and
            treatment test conditions 81
Figure 44. Percent change in LDW-C right alert rates between baseline and
            treatment test conditions 81
Figure 45. Drivers' opinions of the usefulness of the integrated system displays 82
Figure 46. Drivers' opinions of the integrated system auditory warnings 83
Figure 47. Number of mute button uses during the field test... 83
Figure 48. Number of volume adjustments during the field test.. 84
Figure 49. Drivers' opinions of the usefulness of system controls 84
Figure 50. Availability of LDW function by travel speed 85
Figure 51. Block diagram of longitudinal driving conflicts 107
Figure 52. Block diagram of lateral driving conflicts 107

List of Tables

Table 1.	Annual frequency of target crashes by pre-crash scenario	8
Table 2:	Distribution of subject heavy trucks by pre-crash scenario and speed limit	9
Table 3.	Demographics of field test participants	11
Table 4.	Exposure of test subjects in the field test	12
Table 5.	Heavy-truck video sampling rates	14
Table 6.	Results of baseline versus treatment paired t-test for average speed in mph	19
Table 7.	Results of baseline versus treatment paired t-test for mean headway in seconds	20
Table 8.	Results of baseline versus treatment paired t-test for lane changes per 10 VMT	20
Table 9.	Results of baseline versus treatment paired t-test for percent signaled lane changes	21
Table 10.	Results of baseline versus treatment paired t-test for lane excursions per VMT	22
Table 11.	Results of baseline versus treatment paired t-test for lane excursion duration in seconds	22
Table 12.	Most common secondary task behavior by route type	24
Table 13.	Tasks contributing to the overall change in secondary task engagement, by driver	26
Table 14.	Sign test results for secondary task and eyes-off-forward-scene behavior	28
Table 15.	Average number of driving conflicts per 100 miles driven in baseline versus treatment	30
Table 16.	Average number of driving conflicts per 100 miles driven in baseline versus fourth treatment period	31
Table 17.	Average number of driving conflicts per 100 miles driven in baseline versus treatment in last two hours of the work shift	31
Table 18.	Average measures of driver response to driving conflicts in baseline versus treatment	32
Table 19.	Paired t-test results of average number of specific near-crashes per 1,000 miles driven in baseline versus treatment	37
Table 20.	Paired t-test results of average number of road-departure near-crash types per 1,000 miles driven in baseline versus treatment	38
Table 21.	Driver demographic categories used in driver-acceptance analysis	43
Table 22.	Driver experience categories used in driver-acceptance analysis	44
Table 23.	System features most liked by drivers	45
Table 24.	System characteristics least liked by drivers	46
Table 25.	Number of drivers who found the system to be most helpful in driving situations	46
Table 26.	Changes in driving behavior due to integrated system use	58
Table 27.	Analysis of system alerts	70
Table 28.	Correlation between alerts and driving conflicts/near-crashes	71
Table 29.	Average number of alerts per 100 miles driven in baseline versus treatment	78
Table 30.	Average number of alerts per 100 miles driven in baseline versus treatment for FCW-M, LDW-C left, and LDW-C right alert types	79
Table 31.	Availability of LDW function by travel speed and driving conditions	86
Table 32.	Estimated safety benefits of the integrated system based on alert rate reduction	87
Table 33.	Data mining variables	108
Table 34.	Processed numerical field test data and corresponding system alerts	109
Table 35.	Breakdown of analyzed alert videos	112
Table 36.	Variables coded in video analysis by alert type	113

List of Acronyms

B	Baseline
Ax_{avg}	Average deceleration level during conflict resolution
Ax_{min}	Minimum deceleration level during conflict resolution
Ay_{max}	Maximum lateral acceleration
BSM	Blind spot monitor
CDL	Commercial driver's license
DAS	Data acquisition system
dLE_{max}	Maximum lane excursion distance
DVI	Driver-vehicle interface
ER	Exposure ratio
FCW	Forward crash warning
FCW-M	Forward crash warning for moving vehicles
FCW-S	Forward crash warning for stationary objects
FOT	Field operational test
GES	General Estimates System
IVBSS	Integrated Vehicle-Based Safety Systems
LCM	Lane-change/merge
LDW	Lane-departure warning
LDW-C	Cautionary lane-departure warning
LDW-I	Imminent lane-departure warning
LH	Line-haul
LVD	Lead vehicle decelerating
LVM	Lead vehicle moving
LVS	Lead vehicle stopped
MDAT	Multimedia data analysis tool
P&D	Pick-up and delivery
PR	Prevention ratio
SQL	Structured Query Language
T	Treatment
T_4	Fourth period of treatment condition
tLE	Duration of lane excursion
TTC_B	Time-to-collision at brake onset
TTC_{min}	Minimum time-to-collision during conflict resolution
UMTRI	University of Michigan Transportation Research Institute
VMT	Vehicle miles traveled

Executive Summary

Introduction

In November 2005, the U.S. Department of Transportation entered into a multi-year cooperative research agreement with an industry team led by the University of Michigan Transportation Research Institute to develop and test an integrated, vehicle-based crash warning system that addressed rear-end, lane-change and roadway departure crashes for light vehicles and heavy commercial trucks. The work carried out under this agreement was known as the Integrated Vehicle-Based Safety Systems program. The 5-year effort was divided into two consecutive, non-overlapping phases.

The UMTRI-led team was responsible for the design, build, and field-testing of the prototype integrated crash-warning systems. The heavy-truck platform team included Eaton Corporation, Navistar, Takata Corporation, Con-way Freight, and Battelle.

The IVBSS program team also included senior technical staff from the National Highway Traffic Safety Administration, the Federal Motor Carrier Safety Administration, the Research and Innovative Technology Administration, the National Institute for Standards and Technology, and the Volpe National Transportation Systems Center. RITA's Intelligent Transportation Systems Joint Program Office was the program sponsor, providing funding, oversight, and coordination with other U.S. DOT programs. The cooperative agreement was managed and administered by NHTSA, and the Volpe Center acted as the program independent evaluator.

This report presents the methodology and results from the independent evaluation of the heavy-truck field operational test.

Background

The purpose of the IVBSS program is to assess the potential safety benefits and driver acceptance associated with a prototype integrated crash warning system designed to address rear-end, roadway departure, and lane change/merge crashes for light vehicles and heavy commercial trucks. The heavy-truck integrated system included the following types of crash- imminent warnings:
- Forward crash warning (FCW);
    - FCW stopped (FCW-S): refers to alerts issued for stationary objects located in the vehicle's forward travel lane;
    - FCW slower (FCW-M): refers to alerts issued for moving objects, such as a slower moving or a decelerating lead vehicle;
- Lane-change/merge (LCM) warning;

- Lane-departure warning (LDW);
    - LDW cautionary (LDW-C): refers to alerts issued when the vehicle is drifting out of its lane into a clear area (unoccupied lane or clear shoulder); and
    - LDW imminent (LDW-I): refers to alerts issued when the vehicle is drifting into an occupied lane or toward a roadside object, causing potential for a collision.

Evaluation Goals

The goals of the evaluation were to:
- **Achieve a detailed understanding of the system's safety benefits** by estimating the number of crashes that might be avoided from full deployment of integrated safety systems in the U. S. commercial heavy-truck fleet. This goal also considered unintended consequences resulting from changes in driver behavior that could have negative side-effects on traffic safety.
- **Determine driver acceptance** by assessing ease of use, perceived usefulness, ease of learning, drivers' advocacy, and drivers' assessment of their own driving performance while using the integrated safety system.
- **Characterize system performance** by examining the operational performance of the integrated system and its components in the driving environment.

Procedure

The evaluation was based on naturalistic driving data collected from 18 volunteer drivers who operated 10 commercial trucks that were equipped with a prototype integrated safety system over a 10-month period. The driver pool included 8 pick-up and delivery and 10 line-haul drivers with safe driving records. Baseline data was collected during the first 2 months of the test, while the last 8 months included collection of driver performance data with the system enabled.

The analysis was performed on data collected from 87,730 miles driven during the baseline period and 409,656 miles driven during the treatment period. In addition to numerical data analysis, a sample of 14,405 videos corresponding to system alerts was selected for detailed examination including 6,314 alert videos from pick-up and delivery and 8,091 alert videos from line-haul drivers.

Results

Safety Impact:

- Full deployment of integrated safety systems in the U. S. heavy-truck fleet could prevent between 3,000 and 13,000 police-reported target crashes annually (2% to 11%). This estimate is based on observed reductions in alert rates from forward-crash warnings for moving targets and cautionary lateral-drift warnings between the baseline and treatment periods. The breakdown of this estimate by alert type and pre-crash scenario is shown in the table below.

Function	Pre Crash Scenario	Annual Target Crashes	Maximum Estimated Crash Reduction	Maximum Estimated Effectiveness
FCW-M	Rear end/Lead vehicle decelerating Rear end/Lead vehicle moving	18,000	5,000	27%
FCW-S	Rear end/Lead vehicle stopped	19,000		
LCM	Changing lanes/same direction Turning/same direction	53,000	Insufficient data to estimate	
LDW-I	Drifting/same lane	7,000		
LDW-C Left	Opposite direction/No maneuver Road edge departure/No maneuver	11,000	3,000	29%
LDW-C Right	Road edge departure/No maneuver	15,000	5,000	36%
Integrated System	All	123,000	13,000	11%

- Drivers experienced an 11-percent drop in road departure near-crashes with the system enabled, especially to the left side, indicating that the alerts improved driver awareness of the position of their vehicle in their travel lane.
- The number of unintentional crossings of both lane boundaries was reduced with the system enabled. There was a 9-percent reduction at speeds between 35 and 55 mph, and by 15 percent at speeds over 55 mph, suggesting that lateral-drift warnings helped drivers maintain better lane position.
- Line-haul drivers increased their turn signal use by 5 percent when changing lanes at speeds over 45 mph with the system enabled.
- Improvements in turn-signal use and lane-keeping performance continued into the last 2 months of the treatment period, indicating lasting effects of system use.
- Eight line-haul drivers were involved in more secondary tasks (eating, drinking, and talking on a cellular phone) with the system enabled, but no negative effects on driving performance could be associated with this increase.
- Eight drivers reported that the integrated system helped them avoid a crash or near-crash.

- Drivers experienced lower lateral-drift cautionary alert rates with the system enabled for left- and right-lane excursions (21% and 17%, respectively), indicating that the integrated system helped them stay within their lane.
- Forward-crash alert rates for moving targets were 12 percent lower with the system enabled, suggesting improved driver awareness in response to vehicles ahead of them.

Driver Acceptance:
- Drivers found the system to be easy to learn and use, and found the auditory alerts easy to understand.
- Fifteen of the 18 drivers said that they would prefer driving a truck with the integrated system over a conventional, unequipped truck.
- Fifteen drivers felt that the system increased their situational awareness.
- Thirteen drivers indicated that they felt that driving with the system would increase their driving safety.
- Fifteen drivers said that they would recommend that their employer purchase the integrated system for their vehicle fleet.

System Performance:
- Ninety percent of lateral-drift cautionary alerts were issued when drivers crossed lane boundaries without using their turn signals.
- The forward crash warning system consistently misclassified stationary objects as in-path targets. For example, 97 percent of all warnings issued for stationary targets were for out-of-path objects, e.g., overhead bridges and signs and roadside objects, while only 7 percent of alerts issued for forward-crash moving targets were for out-of-path objects.
- More than half of all lane change/merge-imminent alerts were issued when no target was present in the adjacent lanes. The low reliability of these warnings can be attributed to false targets from radar reflections off the trailer body (Sayer et al., 2010).
- The lateral-drift warning function met its lane-tracking performance requirements for all speed ranges.

Conclusions

Overall, driving with the integrated system improved driver's performance by increasing their awareness of traffic around them and the position of their vehicle in their travel lane. The system encouraged better lane-keeping behavior by alerting them when they were drifting out of their lane, and helped avoid potential rear-end crashes by letting them know when they were closing in or approaching a lead vehicle too closely. These features increased driver's awareness of their driving habits and helped improve their vigilance. Lateral-drift cautionary alerts reminded drivers to use their turn signal more regularly, a habit conducive to safe driving.

Drivers who participated in the study had favorable opinions of the system and thought that it would improve their driving safety. Since safety was a company-wide priority, almost all drivers said they would prefer driving a truck with the integrated safety system than a standard, unequipped truck. Most drivers were aware of the system's shortcomings and reported receiving false warnings and being annoyed by them, at least on occasion. Despite the system's shortcomings, drivers maintained their favorable view and desire to use the integrated system as a means of increasing their driving safety.

Poor reliability of side object classifiction and consistent misclassification of stopped objects in the vehicle's forward path were two shortcomings of the prototype system in need of improvement.

1. Introduction

This report presents the analytical approach and results of the independent evaluation of a prototype integrated crash warning system for heavy trucks. The evaluation is based on naturalistic driving data collected from a commercial fleet of 10 heavy trucks equipped with a prototype integrated safety system. The analytical methods used in the evaluation are outlined, and results are presented and discussed.

> **HIGHLIGHTS**
> - The safety system combines rear-end, lane-change, and lane-departure crash warning functions that address approximately 194,000 police-reported crashes involving medium and heavy trucks annually.
> - 18 professional drivers on 10 line-haul and 8 pick-up and delivery routes accumulated over 600,000 miles driving 10 heavy trucks equipped with the integrated safety system over a 10-month period.
> - Approximately 87,000 alerts were issued during an 8-month treatment period while the system was enabled.

1.1 Integrated Safety System

The integrated safety system provides information to assist drivers in avoiding or reducing the severity of crashes in which an equipped truck:

- Strikes the rear end of another vehicle (FCW);
- Changes lanes, initiates a turn and encroaches on another vehicle in an adjacent lane, or merges into traffic and collides with another same-direction vehicle (LCM); and
- Unintentionally drifts off the road edge or crosses a lane boundary (LDW).

The integrated safety system for heavy trucks consists of three primary crash warning functions (Sayer et al., 2009):

- Forward crash warning (FCW):
  - FCW stopped (FCW-S): alerts issued for a stationary object
  - FCW slower (FCW-M): alerts issued for moving objects, such as a slower moving or decelerating lead vehicle
- Lane-change/merge (LCM) warning
- Lane-departure warning (LDW):
  - LDW cautionary (LDW-C): alerts issued when the truck is drifting out of its lane into a clear area (unoccupied lane or clear shoulder)
  - LDW imminent (LDW-I): alerts issued when the truck is drifting into an occupied lane or towards a roadside object, causing potential for a collision

The driver-vehicle interface consists of visual and audio alerts that warn the driver of the occurrence of one of the above situations. Each alert has a unique audio tone and message shown on a center display to assist the driver in understanding which type of threat is present. The heavy truck is also equipped with blind spot monitors to help increase driver awareness of objects that are in the driver's blind spots. The LDW-C and LDW-I alerts provide audio and visual warnings to the driver. Figure 1 shows the center display and blind spot monitors mounted in the heavy-truck cabin.

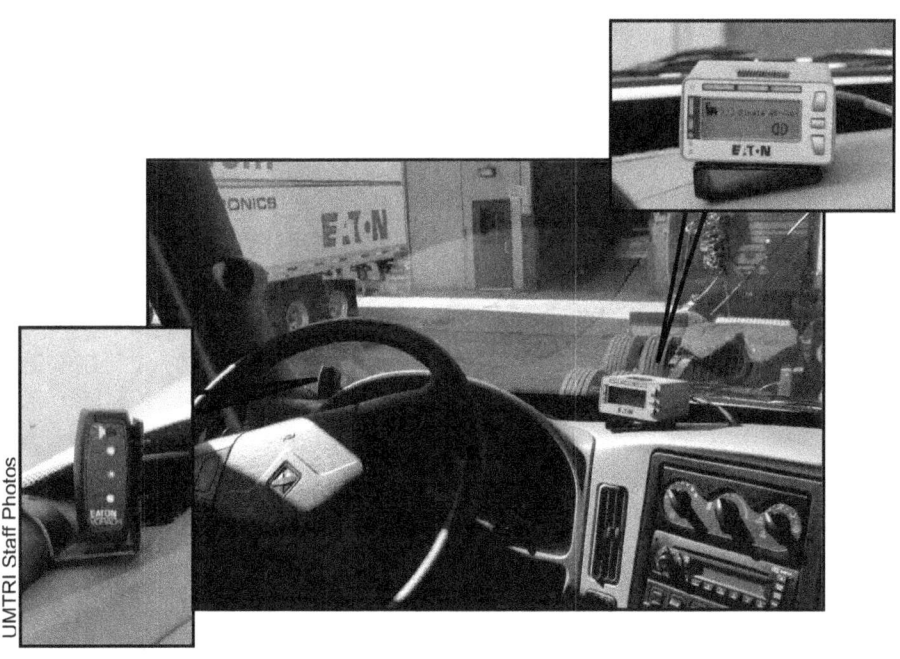

Figure 1. Driver-vehicle interface of the integrated safety system

1.2 Target Crashes

The integrated safety system was designed to address the following pre-crash scenarios, which identify vehicle movements and the critical event prior to a crash (Najm et al., 2007):

- Changing lanes—same direction: vehicle is changing lanes, passing, or merging and then encroaches into the lane of another vehicle that is traveling in the same direction.
- Turning—same direction: vehicle is turning left or right at a junction and then cuts across the path of another vehicle initially going straight in the same direction.
- Drifting—same lane: vehicle is going straight or negotiating a curve and then drifts into the lane of an adjacent vehicle traveling in the same direction.
- Rear-end—lead vehicle stopped (LVS): vehicle is going straight and then closes in on a stopped lead vehicle. Vehicle may also be decelerating or starting in traffic and closes in on a stopped lead vehicle. In some of these crashes, the lead vehicle first decelerates to a stop and is then struck by the following vehicle. This typically happens in the presence of a traffic-control device or when the lead vehicle is slowing down to turn.
- Rear-end—lead vehicle decelerating (LVD): vehicle is going straight while following another lead vehicle and then the lead vehicle suddenly decelerates. Vehicle may also be decelerating in traffic and then closes in on a decelerating lead vehicle.
- Rear-end—lead vehicle moving (LVM): vehicle is going straight or decelerating in traffic and then closes in on a lead vehicle moving at a lower constant speed.
- Road-edge departure—no maneuver: vehicle is going straight or negotiating a curve and then departs the edge of the road at a non-junction area. Vehicle was not making any

maneuver such as passing, parking, turning, changing lanes, merging, or a prior corrective action in response to a previous critical event.
- Opposite direction—no maneuver: vehicle is going straight or negotiating a curve and then drifts and encroaches into the lane of another vehicle traveling in the opposite direction.

Based on crash statistics from the 2004-2008 General Estimates System (GES) crash databases, heavy trucks (gross vehicle weight rating over 10,000 pounds) were involved in crashes preceded by these eight pre-crash scenarios at an average annual frequency of about 194,000 police-reported crashes. In these crashes, approximately 123,000 heavy trucks were the subject vehicle making a maneuver (i.e., changing lanes, merging, turning, or drifting) or following a lead vehicle.

Table 1 ranks the target pre-crash scenarios by frequency of heavy-truck involvement as the subject vehicle. The LCM function addresses changing lanes and turning pre-crash scenarios that accounted for 43 percent of all subject heavy trucks. The FCW function deals with rear-end pre-crash scenarios that were associated with 30 percent of all subject heavy trucks. The LDW function addresses the remaining 27 percent of subject heavy trucks that drifted out-of-lane, resulting in road-edge departure, opposite-direction crash, or same-direction crash. The LDW-C function addresses road-edge departure and opposite-direction pre-crash scenarios, whereas the LDW-I function focuses on heavy trucks in drifting or same direction pre-crash scenarios.

Table 1. Annual frequency of target crashes by pre-crash scenario

Pre-Crash Scenario	Crashes	Heavy Trucks	% Heavy Trucks
Changing lanes/same direction	51,000	29,000	23.6%
Turning/same direction	28,000	24,000	19.5%
Rear-end/lead vehicle stopped	32,000	19,000	15.4%
Road edge departure/no maneuver	18,000	18,000	14.6%
Rear-end/lead vehicle decelerating	18,000	11,000	8.9%
Opposite direction/no maneuver	13,000	8,000	6.5%
Drifting/same lane	20,000	7,000	5.7%
Rear-end/lead vehicle moving	14,000	7,000	5.7%
Total	194,000	123,000	100.0%

Table 2 provides annual crash statistics describing the distribution of heavy trucks in each target pre-crash scenario by the posted speed limit based on average values from the 2004-2008 GES crash databases. The posted speed limit serves as a surrogate measure of travel speed. As seen in Table 2, over 50 percent of the subject heavy trucks were traveling on roadways with posted speed limits of 55 mph or higher when involved in changing lanes/same direction, rear-end/LVD, opposite-direction/no-maneuver, and rear-end/LVM pre-crash scenarios. On the other hand, over 50 percent of the subject heavy trucks were traveling on roadways with posted speed limit under 55 mph when involved in turning/same direction, rear-end/LVS, road-edge departure/no-maneuver, and drifting/same-lane pre-crash scenarios.

Table 2. Distribution of subject heavy trucks by pre-crash scenario and speed limit

Pre-Crash Scenario	Heavy Trucks	Speed Limit (mph)		
		0-20	25-50	55-75
Changing lanes/same direction	29,000	0%	37%	62%
Turning/same direction	24,000	3%	82%	15%
Rear-end/lead vehicle stopped	19,000	1%	66%	33%
Road edge departure/no maneuver	18,000	4%	64%	32%
Rear-end/lead vehicle decelerating	11,000	0%	47%	53%
Opposite direction/no maneuver	8,000	0%	40%	59%
Drifting/same lane	7,000	1%	49%	50%
Rear-end/lead vehicle moving	7,000	0%	34%	66%

1.3 Field Operational Test

The FOT included 20 drivers from Con-way Freight, Inc. who drove 10 tractors on pick-up and delivery and line-haul routes. Pick-up and delivery drivers worked during the day, making many short trips throughout the metropolitan Detroit area to pick up and deliver goods. Line-haul drivers worked the night shift, generally making one long round-trip delivery. Each truck was assigned to 1 pick-up and delivery and 1 line-haul driver. The test participants drove the instrumented vehicle on their normal work route; business operations were not altered in any way during the field test. Participation in the field test was offered to all Con-way drivers on a voluntary basis and drivers were compensated for their participation.

The field test started in February 2009 and was completed on December 14, 2009. The experimental design of the test was an AB design, meaning that each subject experienced two test conditions over a period of 10 months. During the first condition (**A**B), called the baseline period, subjects drove the equipped truck for about 2 months with the integrated safety system turned off. In the second condition (A**B**), or treatment period, subjects drove the truck for about 8 months with the integrated safety system enabled. Even though the system was disabled during the baseline period, the on-board data acquisition system (DAS) recorded all alerts.

While 20 drivers were initially recruited for the field test, data from only 18 drivers were used in this analysis. Due to a combination of economic factors and personal reasons, 2 of the pick-up and delivery drivers did not accumulate sufficient mileage for analysis, resulting in a group of 8 pick-up and delivery drivers and 10 line-haul drivers.

Each driver completed two survey forms and participated in a debriefing interview. Prior to participating in the field test, drivers completed surveys to collect demographic data and information about their driving history. At the end of their participation in the field test, each driver completed a post-drive survey that contained broad categories to measure overall attitudes towards the integrated safety system, as well as items related to driver acceptance. Most items on the post-drive survey asked drivers to rate various items on a 7-point scale with anchored points ranging from strongly disagree to strongly agree. Survey response types also included yes-no and open-ended questions. Appendix A provides an example of the post-drive survey used. Drivers spent approximately 30-45 minutes completing the survey and then reviewed their answers with a researcher to ensure that all sections had been completed correctly, clarify responses, and give drivers an opportunity to discuss any area of interest to them.

1.3.1 Demographics of Field Test Participants

A summary of demographic information for the 18 test participants is shown in Table 3. The participants were all male and ranged in age from 32 to 63 years, with an average age of 47 years. All drivers had a commercial driver's license (CDL) for over 10 years, with an average time of 21.7 years. All drivers had been employed by Con-way Freight, Inc. for a minimum of 8 years, with one driver having been with the company for 25 years. On average, line-haul drivers were slightly older, had held a CDL for more years, and had been with Con-way longer than the pick-up and delivery drivers. Since drivers of line-haul routes are paid more (pay is based on mileage and the line-haul routes are longer than pick-up and delivery routes), these routes are generally driven by drivers with more seniority. A total of 17 out of 18 drivers held high school diplomas, with 2 of these drivers having earned some college credits.

Table 3. Demographics of field test participants

	Driver Number	Age	Years with CDL	Years Employed by Con-Way	Education Level
Pick-up and delivery	1	46	22	15	High School
	2	46	13	10	High School
	4	32	10	10	High School
	5	43	21	10	High School
	6	44	12	9	High School
	7	52	33	8	High School
	8	38	14	11	High School
	10	63	15	14	High School
Line-haul	21	51	30	25	High School
	22	48	27	24	High School
	23	54	35	21	High School
	24	48	25	20	High School
	25	45	21	15	Some College
	26	52	23	18	High School
	27	49	25	19	High School
	28	46	25	12	Some College
	29	53	?	9	11th Grade
	30	40	18	10	High School
	Average P&D	45.5	17.5	10.9	
	Average LH	48.6	25.4	17.3	
	Average All	47.2	21.7	14.4	

1.3.2 Summary of Field Test Exposure

Table 4 presents statistics on mileage driven and experience with system alert types for pick-up and delivery and line-haul drivers (Sayer et al., 2010). On average, pick-up and delivery drivers accumulated 1,851 miles during baseline driving and 7,591 miles in the treatment period, while line-haul drivers accrued 11,452 miles in the baseline period and 41,178 miles in treatment. On average, 470 alerts were recorded for pick-up and delivery drivers during baseline driving, while 1,867 alerts were issued during the treatment period. A larger number of alerts were recorded for line-haul drivers during baseline driving (2,146) and the treatment period (7,171). Pick-up and delivery drivers averaged 26.3 alerts per 100 miles in the baseline period and 23.7 alerts per 100 miles during treatment. For line-haul drivers, the average alert rates per 100 miles were 18.7 during baseline and 16.7 during the treatment period. For all drivers, the number of alerts recorded during the baseline and treatment periods averaged 22.1 and 19.8 alerts per 100 miles, respectively.

Table 4. Exposure of test subjects in the field test

Driver No.	Baseline					Treatment				
	Miles	FCW	LCM	LDW-I	LDW-C	Miles	FCW	LCM	LDW-I	LDW-C
1	1,532	160	44	117	176	9,004	1,007	233	815	1,081
2	2,184	120	98	146	56	7,924	324	455	625	112
4	1,702	137	30	149	47	8,845	877	254	874	173
5	2,387	109	60	170	111	8,919	462	169	515	305
6	2,017	80	75	125	70	2,721	103	97	242	74
7	1,879	152	23	164	228	9,724	718	113	563	919
8	1,527	90	93	211	46	5,461	246	176	313	146
10	1,579	120	78	233	239	8,128	485	205	1,107	1,146
21	10,213	246	82	915	1,253	17,124	295	131	1,515	2,078
22	17,218	277	163	1,447	2,521	66,373	2,310	686	6,348	6,952
23	13,773	330	363	1,267	873	54,539	1,010	1,076	5,731	4,974
24	14,656	503	514	862	647	60,200	3,555	1,872	3,226	2,325
25	14,673	322	266	336	297	18,096	488	384	470	314
26	11,602	214	288	788	139	69,311	1,191	1,583	4,067	731
27	13,078	303	338	1,526	665	58,478	804	1,086	4,189	1,748
28	6,738	118	163	966	636	30,267	697	1,013	3,530	1,145
29	5,858	95	48	284	364	21,968	581	221	938	769
30	6,709	148	66	268	563	15,422	267	246	598	561
Total	129,327	3,524	2,792	9,974	8,931	472,503	15,420	10,000	35,666	25,553

Two crashes occurred during the field test, both during the treatment period. One was a low-speed event in which a pick-up and delivery driver came into contact with a sport utility vehicle while making a wide right turn. The other crash occurred on a line-haul route when the truck struck a deer at night. Neither crash event produced a system alert, since the integrated safety system was not designed to issue alerts at very low travel speeds or for detecting animals at short range and high closing speeds.

1.4 Independent Evaluation

The IVBSS independent evaluation had the following goals (Najm et al., 2006):

- *Achieve a detailed understanding of system safety benefits:* Estimates the number of crashes that could be avoided by the full deployment of the integrated safety system in the commercial heavy-truck fleets in the United States. This goal also addresses unintended consequences in terms of changes in driver behavior that could have negative side effects on traffic safety.
- *Determine driver acceptance of the system*: Assesses the ease of use, perceived usefulness, ease of learning, drivers' advocacy, and drivers' assessment of their own driving performance with the integrated safety system.

- *Characterize system performance*: Examines the operational performance of the integrated safety system and its components in the driving environment.

1.4.1 Data Processing

Data analysis in the independent evaluation involved many forms of data and data processing procedures. The raw field test data underwent a significant amount of processing in order to synchronize the video with numerical data and to conduct data mining and analysis. Figure 2 presents a flowchart showing each type of data and the data processing procedures. The blocks on the far left of Figure 2 (UMTRI data, video data, and numerical data) represent the raw field data. The blocks at the far right end of the figure (video processing, data mining, data logger, and data viewer) refer to the data types and processes created by the independent evaluator; the lowest block (data tables) represents the process output. More detailed information on the data and video processing procedures used to conduct this analysis can be found in Appendix B.

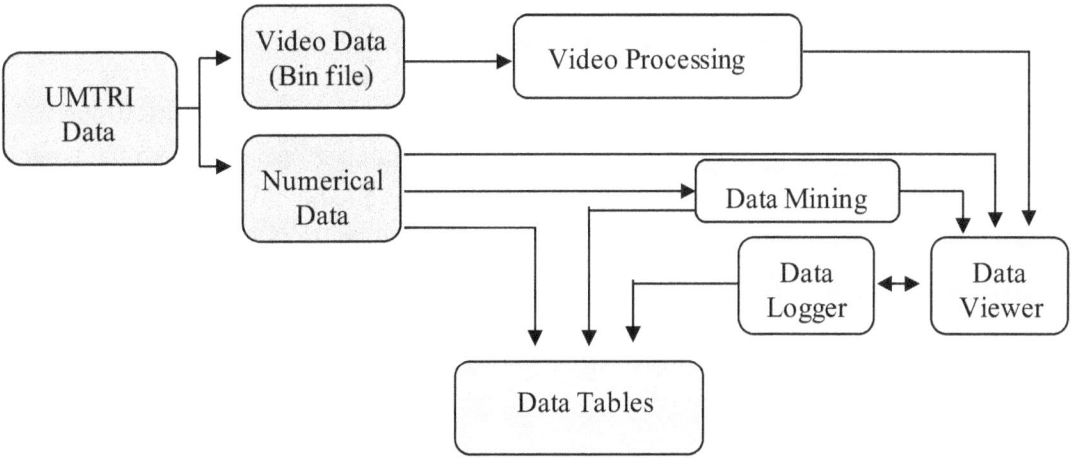

Figure 2. Data processing procedures

The raw video data consist of *.bin* files. Each trip had video files recorded from five different video cameras at different sampling rates, as shown in Table 5. The video processing block in Figure 2 represents the process used to convert the raw video files into a format that allows each file to be synchronized with the numerical data and be compatible with the Volpe Center's custom data analysis tool. The first step involved in video processing was to convert the binary video files into standard *.avi* video format. The second step involved is recompression of the *.avi* files to remove any corrupt frames or errors. This conversion and recompression process allows synchronization of the video data at different sampling rates, frame by frame, and with the numerical data by creating a mapping from each numerical data point to the corresponding frame in the video data. This level of synchronization is necessary to extract certain information about system performance.

Table 5. Heavy-truck video sampling rates

Camera Type	Sampling Rate
Forward view	5 Hz
Driver's face	5 Hz
Cabin/instrument panel	2 Hz
Left side of truck	2 Hz
Right side of truck	2 Hz

The raw numerical data was stored in a Structured Query Language (SQL) database format, and consisted of 10 Hz and 100 Hz data. The raw data is processed by data mining algorithms and is synchronized with video data so that it can be viewed directly. The data mining block in Figure 2 represents the process by which the data mining algorithms are run on the raw numerical data to produce tables of new variables stored in a separate database.

Once all video had been processed and synchronized with the numerical data, the Volpe Center's data analysis tool was used to extract information about system performance from the videos. This method allowed the analyst access to objective information about the driving scene (e.g., speed, distance to lead vehicle, turn signal usage) as a supplement to the video. As the video is reviewed, objective information is extracted and entered into the data logger and then stored in a numerical database. The results of the data mining algorithms and video analysis, as well as some of the raw numerical data, are then extracted using SQL queries. The data tables, shown at the bottom of Figure 2, were used to conduct all analyses.

1.4.2 Multimedia Data Analysis Tool

The Volpe Center developed a multimedia data analysis tool (MDAT) to extract objective information from the five video data channels collected during the field test. While the numerical data provide information about vehicle dynamics and the driving scenario, some information can only be obtained from examining the video. Video analysis is used to supplement the numerical data.

The MDAT is used to synchronize and simultaneously play back five video channels, presenting a full view of the driving scene and driver. In addition to video data, the MDAT is connected directly to the numerical database and can display any of approximately 200 numerical data channels along with the video. Synchronizing the video with numerical data allows the viewer full access to all of the information necessary to fully assess the driving scenario and driver condition.

Figure 3 shows a screen view of the MDAT. The left side of the viewing window shows five channels of video data: front road scene, driver face, cabin, left-side road scene, and right-side road scene. The video is controlled by the buttons on the bottom of the window and the numerical data can be displayed in a separate window. Drop-down menus on the right side of the screen are provided to code specific information about the video as viewers watch videos of interest. The information entered in these menus is saved in a table as part of the field test database, making it accessible to support further analysis. In this analysis, a sample of 14,405 heavy-truck alerts (6,286 pick-up and delivery and 8,090 line-haul) were viewed and coded. Detailed information about the video sampling and video analysis can be found in Appendix D and Appendix E.

Figure 3. Screen view of multimedia data analysis tool

2. Safety Impact

This analysis addresses the safety benefits goal of the independent evaluation by asking two key questions:

- If all heavy trucks in the U.S. vehicle fleet were equipped with the integrated safety system, what would be the annual change in the total number of rear-end, lane-change, and road-departure crashes?
- Would use of the integrated safety system result in unintended consequences that might impact overall traffic safety in a negative or positive manner?

The first question deals with the estimation of potential safety benefits that would result from full deployment of integrated safety systems. The second question looks for any unintended driving behavior from system use that could potentially cause harm to the equipped truck or other road users.

> **HIGHLIGHTS**
> - Line-haul drivers increased their turn signal use from 78 percent of lane changes during baseline to 83 percent with the system enabled.
> - With the system enabled, drivers showed a reduction in unintentional lane crossings.
> - Nine line-haul drivers were involved in more secondary tasks with the system enabled.
> - Drivers experienced and 11 percent drop in road-departure near-crashes with the system enabled.
> - Improvements in turn signal use and lane-keeping continued into the fourth treatment period, indicating lasting effects of system use.

The integrated system was designed as a countermeasure to a number of pre-crash scenarios that occur immediately before rear-end, lane-change, and road-departure crashes (Najm et al., 2007). Safety benefits are derived from the system's effectiveness in reducing the frequency of target pre-crash scenarios listed in Table 1. The LDW function may also prevent opposite-direction crashes due to unintentional drifting into a left-adjacent lane of oncoming traffic.

2.1 Safety Impact Technical Approach

Figure 4 illustrates the analysis framework used to assess safety impact. This framework divides the test subjects' driving experience into three areas: overall experience; driving conflicts; and near-crashes. In general, the safety analysis compares the test subjects' driving experience between the baseline (B) and treatment (T) treatment periods.

Results from the analysis were synthesized to project potential safety benefits. Safety benefits are expressed in terms of the system's potential to reduce the number of target crashes. These benefits are ideally measured from actual crash data; however, only 2 crashes were observed during the conduct of the field test. Thus, this analysis estimates the safety benefits by applying a methodology that uses non-crash, performance data (driver, vehicle and system) collected during the field operational test (Ference et al., 2006).

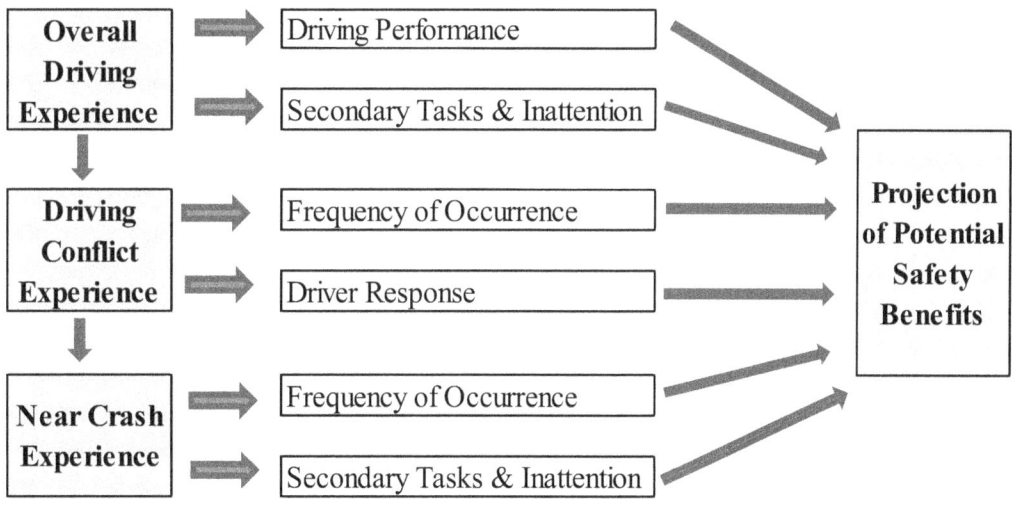

Figure 4. Safety benefits framework

2.2 Overall Driving Experience

This analysis addresses driver performance during the field test. Driving measures are compared "within subjects" for each route type between the baseline and treatment test conditions. In addition to examining any change in driver behavior between test conditions, this analysis looks into long-term adaptation to the system by comparing performance during baseline driving to the fourth period of the treatment (T_4) period. It should be noted that the duration of baseline driving was about 2 months, while the treatment condition averaged six to 8 months. The treatment condition was split into four periods by miles driven, each accounting for approximately one quarter of the miles during the treatment condition. Mileage for the treatment periods varied slightly because individual trips were not assigned to the two different test periods. The fourth treatment period, or T_4, represents the last quarter of each driver's treatment exposure, or their performance during approximately the last 2 months of the test.

Two-tail paired t-tests were performed to compare driver performance. A paired t-test is used to determine if there is a statistically-significant difference between the means of the same subjects under different circumstances. A two-tailed test is used when the mean under the test condition can be either greater than or less than the mean during baseline. For all t-tests conducted in this analysis, a p value of 0.05, or 95 percent confidence, was used to define statistical significance. These values are indicated by bold font in the tables throughout this report.

The measures used in this analysis for driving performance include the following:
- Travel speed (mph): the truck is traveling at constant speeds over 35 mph;
- Time headway (in seconds): the truck is following a lead vehicle, both traveling at constant speeds. This variable is assessed under two speed conditions:

- Travel speeds between 35 mph and 55 mph; and
- Travel speeds above 55 mph.
* Lane change: the truck is traveling at speeds above 45 mph. This maneuver is evaluated using two measures:
  - Number of lane changes per vehicle miles traveled; and
  - Proportion of signaled lane-change maneuvers.
* Lane keeping: assessed using two measures under two speed conditions:
  - Measures (number of lane excursions[1] per vehicle miles traveled, and duration of lane excursion); and
  - Speed conditions (travel speeds between 35 mph and 55 mph, and ravel speeds above 55 mph).

The measures for inattentive behavior include:
* Secondary tasks: driver involvement was assessed using two measures:
  - Proportion of analyzed alerts with secondary tasks; and
  - Number of secondary tasks per analyzed alert.
* Eyes-off-forward-scene: driver scan of the forward scene was evaluated by capturing eyes-off-the-road for a continuous duration of more than 1.5 seconds during a 5-second period prior to onset of alerts with the measure:
  - Proportion of analyzed alerts with eyes-off-forward-scene more than 1.5 seconds.

The analysis of the overall driving experience was conducted in two period comparisons using paired *t*-test for means between:
* Baseline and all treatment periods; and
* Baseline and T4 treatment periods.

The analysis was also performed for three test subject groups:
* Pick-up and delivery drivers;
* Line-haul drivers; and
* All drivers.

2.2.1 Speed Maintenance

The average speed of each driver was calculated for segments in the processed numerical database for speeds greater than 35 mph. Table 6 presents the results of the paired *t*-tests for this dataset. It is clear from this table that a distinct difference exists in the maintained speed between the two driver groups. This is not unexpected, given the differences in driving routes. Line-haul drivers spend most of their time at or near the maximum posted speed limit.

---

[1] Lane excursions refer to a scenario where any of the vehicle's wheels cross the lane line of the lane in which the vehicle is currently traveling while its turn signal is not activated.

There was no effect of the integrated system on the speed pick-up and delivery drivers maintained during the treatment period. The differences between baseline and treatment values were small and the p values did not imply statistical significance. In contrast, there was statistical significance for the line-haul driver group when comparing the baseline period to the entire treatment period. The magnitude of the change–less than one mph–may appear small at first, but when one considers that maximum speed is mechanically regulated, this incremental change toward the maximum indicates a change in behavior indicative of increased driver confidence at maximum speed. Note that the speed of the fleet trucks was governed by the engine controller at 62 mph and might exceed this set maximum speed when trucks are traveling on down grades. The slight increase in speed for line-haul drivers during the treatment period may have been due to weather and traffic conditions, as much of the treatment period took place during the summer months, while the baseline period and T_4 included winter months.

The increase in speed for line-haul drivers in the fourth treatment period was not statistically significant. Nonetheless, the higher average and p value suggest a trend towards an increase in speed in the fourth treatment period.

Table 6. Results of baseline versus treatment paired *t*-test for average speed in mph

Route Type	Baseline Mean	Treatment Mean	p	T4 Mean	p
Pick-up and delivery	42.4	42.5	0.86	42.8	0.58
Line-haul	59.1	59.8	**0.05**	59.7	0.18
All	51.7	52.1	0.12	52.2	0.17

2.2.2 Headway Keeping

The time headway to a lead vehicle was calculated for each driver when the vehicle was traveling between 35 mph and 55 mph, and for speeds greater than 55 mph. Results of the paired *t*-test for this dataset are given in Table 7. There is a noticeable difference between the values of this parameter for the two driver groups in the intermediate speed range. It may well be that the environment of line-haul drivers allows for more lane changes and cut-ins by other drivers, resulting in more difficulty maintaining desired headway.

Pick-up and delivery drivers showed a trend towards an increase in headway keeping during the treatment period. While not meeting the criterion for statistical significance, the p value was relatively low. Line-haul drivers showed little or no change in their headway-keeping behavior between baseline driving and the treatment period. At speeds above 55 mph, both driver groups were most likely traveling on multiple-lane, limited-access highways where time headway could have been significantly affected by the behavior of other drivers. For this road type, there was very little observed difference between driver groups or treatment periods. Understandably, there is no *t*-test result implying statistical significance.

Table 7. Results of baseline versus treatment paired *t*-test for mean headway in seconds

Route Type	Speed (mph)	Baseline Mean	Treatment Mean	p	T4 Mean	p
Pick-up and delivery	$35 \leq v < 55$	3.16	3.27	0.10	3.23	0.20
	$v \geq 55$	2.26	2.30	0.56	2.20	0.60
Line-haul	$35 \leq v < 55$	2.52	2.56	0.51	2.53	0.90
	$v \geq 55$	2.30	2.32	0.72	2.32	0.73
All	$35 \leq v < 55$	2.81	2.88	0.11	2.84	0.45
	$v \geq 55$	2.28	2.31	0.48	2.27	0.77

2.2.3 Lane-Change Behavior

The number of lane changes per 10 vehicle miles traveled (VMT) was calculated for each driver for all segments when the truck was traveling above 45 mph. The results of the paired *t*-tests for this dataset are given in Table 8, which show a difference in behavior between pick-up and delivery and line-haul drivers. Since line-haul drivers spend more time on long stretches of multi-lane highways, their lane changes are less frequent. Pick-up and delivery drivers spend a great deal of time on surface streets and arterials as opposed to limited-access highways, and can be expected to change lanes far more frequently when traveling to their destinations. In fact, there is more than an order of magnitude difference in lane-change frequency. The integrated safety system had some impact on lane-change frequency. Lane-change maneuvers by pick-up and delivery drivers declined by nine percent between the baseline and treatment test conditions. In contrast, line-haul drivers showed no change. For all drivers, there was a statistically significant drop of 7 percent in lane-change maneuvers from baseline to treatment periods. This change was sustained during the fourth treatment period.

Table 8. Results of baseline versus treatment paired *t*-test for lane changes per 10 VMT

Route Type	Speed (mph)	Baseline Mean	Treatment Mean	p	T4 Mean	p
Pick-up and delivery	$v > 45$	4.3	3.9	**0.02**	3.7	0.10
Line-haul	$v > 45$	1.5	1.5	0.60	1.4	0.31
All	$v > 45$	2.7	2.6	**0.03**	2.4	**0.05**

A parameter that more directly measures driver behavior is the percentage of lane changes in which the driver makes signaled lane changes. This was calculated for each driver by dividing the number of lane changes with an active directional signal by the total number of lane changes for all segments when the driver was traveling more than 45 mph. Any lane change that occurred while a driver was using emergency flashers (as might occur when traveling uphill) was counted as a signaled lane change. Table 9 presents the results of the paired *t*-tests for this dataset. While pick-up and delivery driver behavior was statistically unchanged, line-haul drivers exhibited a statistically significant decrease in the number of unsignaled lane changes from baseline to treatment conditions. For line-haul drivers, signaled lane changes increased by

5 percent between the baseline and the entire treatment test conditions and were sustained during the fourth treatment period.

Table 9. Results of baseline versus treatment paired *t*-test for percent signaled lane changes

Route Type	Speed (mph)	Baseline Mean	Treatment Mean	p	T4 Mean	p
Pick-up and delivery	v > 45	84.2%	81.5%	0.17	80.6%	0.13
Line-haul	v > 45	78.4%	82.2%	**0.01**	82.6%	**0.01**
All	v > 45	81.0%	81.9%	0.47	81.9%	0.62

2.2.4 Lane Keeping

Lane keeping is quantified in terms of lane excursions, i.e., partial or incomplete lane changes in which the host truck crosses a lane boundary and then returns to its original lane. Lane excursions were quantified for each driver by occurrence per VMT and duration. The parameters were calculated in two speed ranges, 35 to 55 mph and above 55 mph. Lane excursion rates were also broken down by excursions to the right and left. Table 10 provides the means and p values of the paired *t*-tests associated with lane excursion (greater than 0.1 mile) per mile traveled in these two speed ranges.

All drivers showed a statistically significant reduction in lane excursion in both directions between test conditions of nine percent at lower speeds and 15 percent at higher speeds. These observed reductions were maintained in the fourth treatment period. These reductions were primarily to the right side, as the change in lane excursions to the left was not statistically significant in either speed range. For all drivers, lane excursions to the right decreased by about 15 percent at lower speeds and 19 percent at higher speeds. For all drivers, lane excursions dropped by about 10 per 100 miles from the baseline period to the fourth treatment period at lower speeds.

Pick-up and delivery drivers showed a statistically significant reduction in lane excursions of about 11 percent on both sides, and nine percent for the right side at lower speeds between the baseline period and the treatment period. This reduction was maintained into the fourth treatment period. These drivers showed no significant change in the frequency of lane excursions on both sides and the right side at higher speeds and to the left side in either speed range. On the other hand, line-haul drivers showed significant drops of 24 percent in lane excursions on both sides at higher speeds, and 18 percent to the right at lower speeds. Overall, drivers stayed within their travel lane more often with the integrated system enabled.

Table 10. Results of baseline versus treatment paired *t*-test for lane excursions per VMT

Location	Speed (mph)	Pick-up and delivery					Line-haul					All Drivers				
		B	T	p	T4	p	B	T	p	T4	p	B	T	p	T4	p
Both Sides	$35 \leq v < 55$	0.80	0.71	**0.05**	0.70	**0.02**	0.92	0.85	0.21	0.82	0.07	0.87	0.79	**0.03**	0.77	**0.005**
	$v \geq 55$	0.36	0.35	0.52	0.38	0.71	0.35	0.26	**0.03**	0.25	**0.04**	0.35	0.30	**0.02**	0.31	0.23
Left	$35 \leq v < 55$	0.45	0.39	0.07	0.39	0.14	0.50	0.51	0.87	0.50	0.89	0.48	0.46	0.34	0.45	0.25
	$v \geq 55$	0.13	0.14	0.44	0.17	0.13	0.15	0.12	0.16	0.12	0.15	0.15	0.13	0.33	0.14	0.90
Right	$35 \leq v < 55$	0.34	0.31	**0.04**	0.31	0.07	0.42	0.34	0.05	0.33	**0.02**	0.39	0.33	**0.01**	0.32	**0.005**
	$v \geq 55$	0.22	0.20	0.41	0.21	0.69	0.19	0.14	0.07	0.13	0.09	0.21	0.17	**0.05**	0.16	0.11

Table 11 presents the results of lane excursion duration for the upper speed ranges. The average duration decreased slightly from the baseline to the treatment conditions for all drivers in these speed ranges. However, this observed difference was not statistically significant.

Table 11. Results of baseline versus treatment paired *t*-test for lane excursion duration in seconds

Route Type	Speed (mph)	Baseline Mean	Treatment		T4	
			Mean	p	Mean	p
Pick-up and delivery	$35 \leq v < 55$	3.7	3.8	0.53	3.9	0.30
	$v \geq 55$	3.2	3.1	0.32	3.1	0.79
Line-haul	$35 \leq v < 55$	5.7	5.2	0.48	5.2	0.59
	$v \geq 55$	4.3	4.0	0.38	3.9	0.32
All	$35 \leq v < 55$	4.8	4.6	0.54	4.6	0.74
	$v \geq 55$	3.8	3.6	0.29	3.5	0.29

2.2.5 Attention to Primary Driving Task

This analysis focused on driver attention to the driving task and the forward scene for all alerts analyzed. Secondary tasks and eyes-off-forward-scene events were recorded for each alert, as discussed in Section 2.2.5.1. Appendix E contains a list of distraction behaviors and a definition of "eyes-off-forward-scene" events. The analysis was conducted for each driver, as well as for all drivers, by route type. Driver attention metrics were also broken down by treatment period.

2.2.5.1 Analysis of Secondary Tasks

Secondary tasks include driver behaviors that compete for attention to the primary driving task and could be potentially distracting to the driver. These tasks were identified by viewing the face and cabin cameras of 14,405 analyzed videos. Table 12 shows the 10 most frequent secondary tasks performed by drivers for each route type. The percentages represent the proportion of the 6,286 pick-up and delivery and 8,090 line-haul alerts analyzed in which each behavior was present. The most frequent secondary task for all drivers was grooming. Pick-up and delivery drivers were more likely to be drinking, text messaging or looking at a cell phone than line-haul drivers, and line-haul drivers were more likely to use a Bluetooth headset, have their eyes closed for greater than one second, or sing. Overall, cell-phone-related activities were present in about 10 percent of pick-up and delivery episodes and 11 percent of episodes for line-haul drivers. All drivers used cell phones while driving and 7 drivers used Bluetooth headsets (all line-haul drivers), and 5 drivers smoked while they drove.

Figure 5 illustrates the change in each driver's secondary task behavior from baseline driving and the treatment period. The bars represent changes in episodes in which any secondary tasks were present. Positive values indicate an increase in the proportion of episodes with secondary tasks, while negative values indicate a reduction. Five of 8 pick-up and delivery drivers increased the proportion of events with no secondary tasks between the baseline and treatment periods, but only 2 of 10 line-haul drivers showed a reduction in their proportion of events with secondary tasks. Overall, line-haul drivers showed larger changes both in the reduction and increase of events with secondary tasks. Driver 23 increased his secondary task proportion from 42 percent during baseline driving to 63 percent in the treatment period. The largest reduction was from driver 22, with secondary tasks present in 70 percent of his baseline episodes, and 58 percent of treatment episodes.

Overall, pick-up and delivery drivers reduced their proportion of events with secondary tasks slightly, from 50 percent in baseline to 49 percent during the treatment period. Line-haul drivers increased their proportion from 55 percent in baseline to 61 percent in treatment. The overall mean effect of the system (1-T/B) was -4.8 percent, with a 95 percent confidence level of 6.1 percent. The difference in the means was 3.2 percent which was not statistically significant by a t-test for two independent groups ($p = 0.089$). Although there is a trend in an increase in the number of episodes in which drivers engaged in secondary tasks during the treatment period, there are large individual differences between drivers and therefore no statistical differences across drivers.

While Figure 5 captures the proportion of alert episodes with and without secondary tasks, it does not account for alerts with multiple secondary behaviors. Overall, line-haul drivers had a slightly higher average number of secondary tasks per alert than pick-up and delivery drivers (0.61 secondary behaviors per alert for pick-up and delivery, 0.75 secondary behaviors per alert for line-haul).

Figure 6 illustrates the changes in the average number of secondary tasks per alert for each driver. A reduction in the number of secondary tasks indicates an overall reduction in activity by a driver, and is shown in Figure 6. Three of 8 pick-up and delivery drivers and one line-haul driver showed reductions in secondary tasks.

Table 12. Most common secondary task behavior by route type

Pick-up and delivery		Line-haul	
Grooming	11%	Grooming	13%
Looking to the side/outside car	8%	Talking on/listening to Bluetooth headset	7%
Smoking/lighting cigarettes	7%	Reaching for object in vehicle	7%
Reaching for object in vehicle	7%	Eating	6%
Eating	6%	Eyes closed for greater than one second	6%
Talking on/listening to phone	6%	Looking to the side/outside car	6%
Adjusting controls	4%	Singing/whistling	5%
Drinking	3%	Talking on/listening to phone	4%
Text messaging	2%	Adjusting controls	3%
Reading cell phone	2%	Smoking/lighting cigarettes	3%

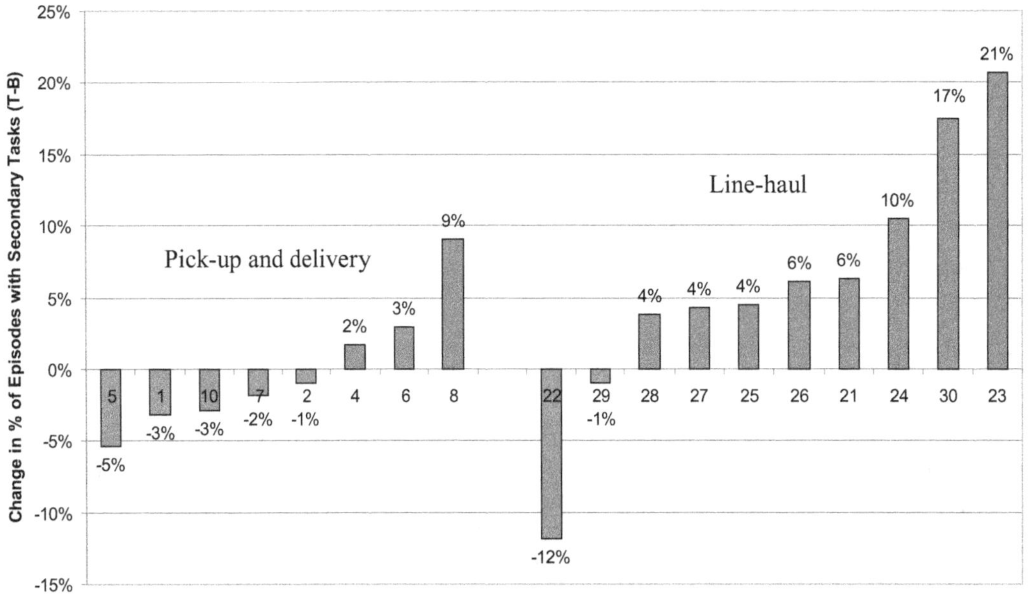

Figure 5. Change in percent of alerts with secondary tasks from baseline to treatment

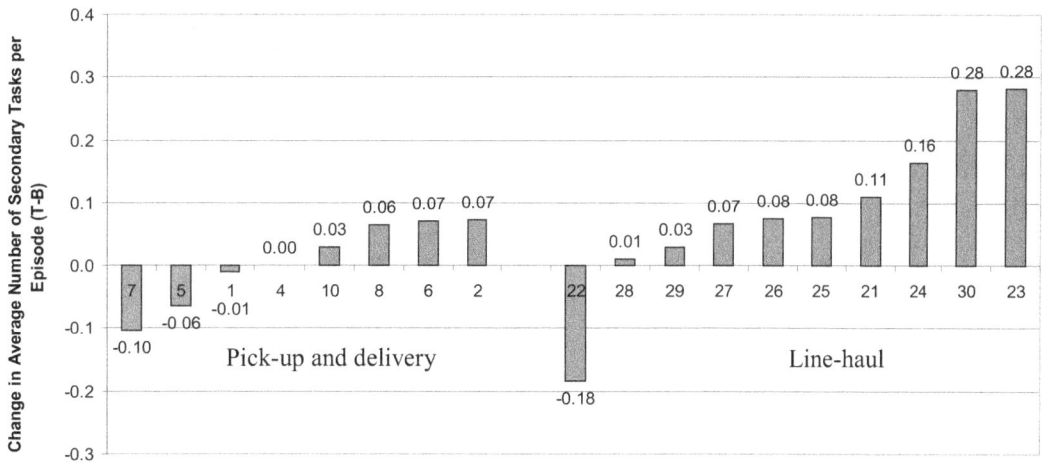

Figure 6. Change in the average number of secondary tasks per alert from baseline to treatment

Results of a paired *t*-test did not show any statistically significant difference in the average number of secondary tasks per alert between the baseline and treatment conditions ($p = 0.06$). The specific tasks that contributed to changes in the average secondary tasks per alert, as shown in Figure 6, varied greatly by driver. There were no overall trends in which tasks increased or decreased across drivers. The tasks that contributed most significantly to the results shown in Figure 6 are listed in Table 13. The positive and negative percentages represent the change in percent of total episodes analyzed that a driver engaged in the task; for example, driver 7 was adjusting controls in 8 percent of all alerts during baseline driving, but only four percent of alerts during the treatment period. The most common tasks that showed a change between baseline and treatment were looking to the side or outside the vehicle (decrease), adjusting controls (decrease), grooming (increase), and talking on a cell phone (increase).

The proportion of viewed alerts with secondary tasks for all drivers increased slightly from baseline to overall treatment and fourth treatment periods. The average value increased from 52.8 percent of the viewed alerts in baseline to 56.1 percent in the overall treatment period and 53.1 percent in the fourth treatment period. This observed difference in driver involvement was not statistically significant based on the two-tail paired *t*-test ($p = 0.09$).

Of the 14,000 videos analyzed, a total of 502 episodes included the secondary task "eyes closed more than one second," an objective metric for capturing instances of drowsy driving. Ninety-four of these episodes occurred during the 2-month baseline period and 408 occurred when the integrated system was enabled (approximately 8 months). Three of the drivers, all line-haul, accounted for over 90 percent of these episodes. One line-haul driver (driver 23), accounted for over half (281) of the episodes and appeared to be drowsy in 19 percent of his baseline alerts and 35 percent of his treatment alerts. Driver 24 accounted for 98 episodes (5% of baseline and 15% of treatment) and driver 23 accounted for 77 episodes (7% of baseline and 12% of treatment).

The other 7 line-haul drivers were not observed to close their eyes regularly or show an increase in the frequency of this behavior between baseline driving and treatment period. The 8 pick-up and delivery drivers accounted for only 5 of the 502 observed episodes where the driver's eyes were closed for more than 1 second. Due to the small number of drivers who exhibited episodes of drowsiness during the field test, no conclusions can be drawn about the effect of the system on drowsiness.

Table 13. Tasks contributing to the overall change in secondary task engagement, by driver

	Driver	Secondary Tasks and Percent Change			
Pick-up and delivery	7	Looking to the side	-9%	Adjusting controls	-4%
	5	Looking outside car	-3%		
	1	Looking outside car	-6%		
	4	Searching interior	-3%	Grooming	+6%
	10	Smoking	+10%		
	8	Grooming	+9%	Talking on phone	+4%
	6	Talking on phone	+5%	Reaching for object	+5%
	2	Smoking	+8%		
Line-haul	22	Looking outside car	-13%	Talking on Bluetooth	-7%
	28	Singing	+9%	Grooming	+5%
	29	Grooming	+8%		
	27	Grooming	+7%	Eating	+6%
	26	Smoking	+9%	Talking on phone	+4%
	25	Looking outside car	+8%	Reaching for objects	+6%
	21	Eating	+12%	Eyes closed	+5%
	24	Eyes closed	+8%	Grooming	+6%
	30	Talking on Bluetooth	+19%	Text messaging	+9%
	23	Eyes closed	+16%		

2.2.5.2 Analysis of Eyes-Off-Forward-Scene

In the video analysis of alert episodes, "eyes-off-forward-scene" was defined as an instance where the driver had his eyes diverted from the forward-driving scene for at least 1.5 continuous seconds in the 5 seconds leading up to the alert. Many drivers showed a pronounced change in the proportion of the eyes-off-forward-scene metric from the baseline to treatment periods. Figure 7 summarizes the changes in the metric from the baseline to the treatment period. Negative values indicate a reduction in the proportion of episodes and positive numbers indicate an increase. Changes ranged from a reduction of 18 percent (29% in baseline and 11% in treatment) for driver 22 to a 17 percent increase (8% in baseline and 25% during the treatment period) for driver 28. Overall, about half the drivers showed an increase in eyes-off-forward-scene behavior from baseline to treatment and half showed a decrease.

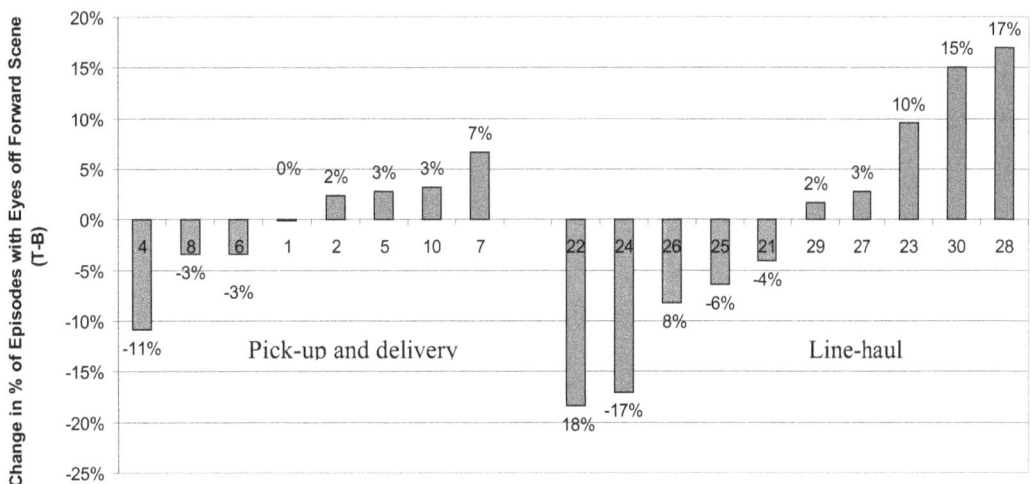

Figure 7. Change in percent of alerts with eyes-off-forward-scene from treatment to baseline

There was no statistically significant difference between the eyes-off-forward-scene percentage between the baseline and the treatment means based on the results of the paired t-test ($p = 0.79$).

The proportion of viewed alerts with eyes-off-forward-scene for all drivers decreased slightly from baseline to overall treatment and fourth treatment periods. The average value dropped from 11.8 percent of the viewed alerts in baseline to 11.2 percent in overall treatment period and to 9.8 percent in the fourth treatment period. This observed difference of drivers' eyes-off-forward-scene was not statistically significant based on the two-tail paired t-test between the baseline and overall treatment periods (p = 0.79) and between baseline and fourth treatment periods (p = 0.37).

Based on the paired t-test, none of the results from driver attention metrics were found to be statistically significant. One of the reasons for the lack of significance could be the large differences between drivers. The large variability between drivers reduces the chances that a paired t-test will show significant results. Since the direction of change, rather than the intensity is of the most interest in this analysis, a nonparametric statistical sign test was used for each driver attention metric.

Table 14 shows the resulting p values from the nonparametric statistical sign test for secondary task and eyes-off-forward-scene behavior. The sign test determines if there is a statistically significant trend towards either an increase or decrease between the baseline and the treatment periods. These results show that line-haul drivers showed an increase in both their percentage of alerts with secondary tasks (8 drivers) and their average number of secondary tasks per alert (9 drivers). Pick-up and delivery and line-haul drivers combined showed a statistically significant increase in the average number of secondary tasks per alert (11 drivers). These results indicate that with the integrated system enabled, drivers engaged in more activities unrelated to driving than when the integrated system was disabled. These secondary tasks may or may not have contributed to driver distraction. As discussed in Sections 2.3 and 2.4 below, there was no statistically significant increase in the number of conflicts or near-crashes with the integrated system enabled. Thus, use of the integrated system does not appear to result in degradation of driving performance, despite a slight increase in involvement in secondary tasks. The sign test did not show a significant trend in the percentage of alert episodes in which the driver's eyes were off the forward scene.

Table 14. Sign test results for secondary task and eyes-off-forward-scene behavior

Route Type	Percentage of Alerts with Secondary Tasks	Secondary Tasks Per Alert	Percentage of Eyes off Forward Scene
Pick-up and delivery	0.67	0.33	0.67
Line-haul	**0.05**	**0.01**	0.64
All	0.23	**0.02**	0.60

2.3 Driving Conflict Experience

This analysis addresses drivers' exposure and response to driving conflicts and examines driving data relevant to safety benefits estimation. As with the overall driving experience analysis, data are compared "within subjects" for each driver type between the baseline and treatment conditions and are grouped by travel speed. Paired t-tests were applied to determine any statistically-significant differences in the mean values of the measures. Additionally, data from the last two hours of each work shift were compared between baseline and treatment conditions.

Driver exposure was assessed by the number of encounters to the driving conflicts per 100 vehicle miles traveled, as listed below:
- Rear-end driving conflicts:
  - Lead vehicle moving and lead vehicle decelerating; and
  - Lead vehicle stopped.
- Lane-change conflicts (it should be noted that analysis of turning conflicts is excluded due to the very low number of encounters with these scenarios)
- Road-departure conflicts combining encounters on straight roads and curves;
- All driving conflicts above combined.

The analysis of driver exposure excludes drivers who were exposed to any conflicts if their total driving mileage under specific conditions is limited to less than 100 miles.

Driver response to each of the four driving conflict types was evaluated using the following measures:
- Rear-end driving conflicts – LVM, LVD and LVS focusing on braking response:
  - Time-to-collision at brake onset (seconds): TTCB;
  - Minimum time-to-collision during conflict resolution (seconds): TTCmin;
  - Minimum deceleration level during conflict resolution (m/s2): Axmin; and
  - Average deceleration level during conflict resolution (m/s2): Axavg.
- Lane-change conflict:
  - Maximum lateral acceleration on straight roads (m/s2): Aymax.
- Road-departure conflict:
  - Maximum lateral acceleration on straight roads (m/s2): Aymax;
  - Maximum lane excursion distance (meters): dLBmax; and
  - Duration of lane excursion(s): tLB.

This analysis also examines the potential effect of the LDW function on opposite direction crashes that involve an unintentional drift into an adjacent lane of opposite direction traffic. This analysis was conducted on a sample of videos capturing driving episodes that resulted in LDW-C or LDW-I alerts being issued. A comparison was performed on driver performance between baseline and all treatment conditions for pick-up and delivery, line-haul, and all drivers using the following measures:
- Proportion of alerts on road edges that do not have adjacent lanes of opposite direction traffic;
- Proportion of alerts for adjacent lanes with opposite direction traffic where there was a vehicle approaching the host truck from the opposite direction; and
- Time-to-collision measured by reviewing the videos from the time of the alert onset until the overlap of the fronts of the two vehicles.

Results of the driving conflict experience, which highlights the differences in driver exposure and response to driving conflicts, are presented below.

2.3.1 Exposure to Driving Conflicts

Table 15 compares the average number of driving conflicts encountered by pick-up and delivery, line-haul, and all drivers per 100 miles traveled between the baseline and treatment test conditions. Driving conflicts are broken down by the combined lead vehicle moving and lead vehicle decelerating, lead vehicle stopped, lane change, and road departure scenarios. Each scenario is also broken down into two speed ranges, (between 25 and 55 mph and speeds above 55 mph). Overall, all drivers were exposed to a higher rate of all driving conflicts in treatment than during baseline driving in two different speed ranges. In individual conflict types, an increase in exposure rate by all drivers was observed in LVM and LVD and lane-change scenarios. On the other hand, a 66-percent decrease in exposure for all drivers was noticed in LVS and road-departure driving conflicts. It should be noted that LVS driving conflicts accounted for only 0.6 percent of all rear-end driving conflicts in all test conditions observed during the field test. All observed differences in driver exposure to driving conflicts between test conditions were not statistically significant.

Long term effects on driver exposure to these scenarios, as measured by the average number of driving conflicts encountered per 100 miles driven, are compared in Table 16. Overall, drivers were exposed to a higher rate of all driving conflicts in the fourth treatment period than during baseline driving at the two different speed ranges. For individual conflict types, an increase in exposure rate for all drivers was observed in all scenarios except for the LVS driving conflict. It should be noted that LVS driving conflicts accounted for only 0.6 percent of all rear-end driving conflicts in the baseline and fourth treatment test conditions combined. As shown in this table, differences in driver exposure to driving conflicts between the baseline and fourth treatment test conditions were not statistically significant.

Table 15. Average number of driving conflicts per 100 miles driven in baseline versus treatment

Driving Conflicts	Speed (mph)	Pick-up and delivery				Line-haul				All Drivers			
		B	T	p	N	B	T	p	N	B	T	p	N
LVM+LVD	$25 \leq v < 55$	13.2	14	0.28	8	1.85	2.16	0.26	10	6.9	7.43	0.13	18
	$v \geq 55$	0.05	0.11	0.54	7	0.003	0.01	0.35	10	0.02	0.05	0.49	17
LVS	$25 \leq v < 55$	0.02	0.03	0.44	8	0.12	0.1	0.88	10	0.07	0.07	0.95	18
	$v \geq 55$	0.73	0.58	0.59	8	1.24	0.42	**0.05**	4	0.9	0.53	0.11	12
Lane Change	$25 \leq v < 55$	0.93	1.21	0.52	8	0.27	0.51	0.08	10	0.56	0.82	0.19	18
	$v \geq 55$	0.77	1.03	0.37	7	0.26	0.3	0.19	10	0.47	0.6	0.25	17
Road Departure	$25 \leq v < 55$	2.69	2.45	0.45	8	2.86	2.42	0.32	10	2.78	2.44	0.2	18
	$v \geq 55$	1.66	1.82	0.79	7	1.23	1.05	0.39	10	1.41	1.37	0.88	17
All Scenarios	$25 \leq v < 55$	16.9	17.7	0.45	8	5.09	5.19	0.88	10	10.3	10.8	0.46	18
	$v \geq 55$	2.48	2.95	0.54	7	1.5	1.36	0.53	10	1.9	2.02	0.72	17

Table 16. Average number of driving conflicts per 100 miles driven in baseline versus fourth treatment period

Driving Conflicts	Speed (mph)	Pick-up and delivery				Line-haul				All Drivers			
		B	T4	p	N	B	T4	p	N	B	T4	p	N
LVM+LVD	25 ≤ v < 55	13.2	14.6	0.21	8	1.85	1.52	0.29	10	6.9	7.33	0.4	18
	v ≥ 55	0.07	0.15	0.64	5	0	0.01	0.23	10	0.03	0.06	0.54	15
LVS	25 ≤ v < 55	0.02	0.01	0.35	8	0.12	0.1	0.88	10	0.07	0.06	0.84	18
	v ≥ 55	0.73	0.5	0.45	8	1.2	0.91	0.65	3	0.86	0.61	0.33	11
Lane Change	25 ≤ v < 55	0.93	1.35	0.4	8	0.27	0.49	0.22	10	0.56	0.88	0.18	18
	v ≥ 55	0.63	1.14	0.22	5	0.26	0.3	0.19	10	0.38	0.61	0.10	15
Road Departure	25 ≤ v < 55	2.69	3.03	0.45	8	2.86	2.83	0.96	10	2.78	2.92	0.68	18
	v ≥ 55	1.56	2.08	0.3	6	1.23	0.98	0.4	10	1.36	1.39	0.89	16
All Scenarios	25 ≤ v < 55	16.9	19	0.18	8	5.09	4.94	0.85	10	10.3	11.2	0.29	18
	v ≥ 55	2.4	3.15	0.32	6	1.5	1.34	0.63	10	1.84	2.02	0.58	16

Table 17 compares the average number of driving conflicts encountered by all drivers per 100 miles traveled between the baseline and treatment test conditions in the last two hours of their work shift. All drivers were generally exposed to a higher rate of all driving conflicts and individual conflict types in the treatment period than during baseline driving in the last two hours of the work shift except for road-departure driving conflicts. Overall, these observed differences in driver exposure during the last two hours of the work shift were not statistically significant based on paired *t*-tests.

Table 17. Average number of driving conflicts per 100 miles driven in baseline versus treatment in last two hours of the work shift

Driving Conflicts	Pick-up and delivery				Line-haul				All Drivers			
	B	T	p	N	B	T	p	N	B	T	p	N
LVM+LVD	18.9	22	0.19	8	0.49	0.36	0.16	10	8.68	9.98	0.2	18
LVS	0.09	0.11	0.72	8	0.01	0.01	0.62	10	0.04	0.05	0.64	18
Lane Change	0.58	1.13	0.13	8	0.32	0.43	0.12	10	0.44	0.74	0.06	18
Road Departure	1.81	1.92	0.81	8	1.81	1.64	0.55	10	1.81	1.76	0.85	18
All Scenarios	21.4	25.2	0.10	8	2.62	2.43	0.57	10	11.0	12.5	0.13	18

2.3.2 Driver Response to Driving Conflicts

Table 18 compares driver response to conflicts between baseline driving and all treatment periods, as well as between baseline driving and the fourth treatment test condition. The results are presented for driver response to three driving conflict types in lower and higher speed ranges. The data was sparse in the higher speed range (over 55 mph) for the three performance measures in the LVM and LVD rear-end driving conflicts. Moreover, the data for the lane-change performance measure were limited at higher speeds for pick-up and delivery drivers and at lower speeds for line-haul drivers. Scenarios in which there were insufficient data to conduct a valid statistical analysis are indicated by the blank cells in Table 18.

Except for two cases, all observed differences in the mean values of the various response measures between test conditions were not statistically significant. There was a statistically significant slight decrease in the mean value of the lateral acceleration applied by line-haul drivers in response to lane-change conflicts at higher speeds (above 55 mph) from 1.4 m/s^2 in baseline to 1.3 m/s^2 in the fourth treatment condition; however, this result was based on only four line-haul drivers. The other statistically significant difference was observed in the mean value of the maximum lane excursion distance in response to road-departure conflicts by all drivers at lower speeds, changing from 0.6 m in baseline to 0.8 m in treatment.

Table 18. Average measures of driver response to driving conflicts in baseline versus treatment

Driving Conflict	Response Measure	Speed Bin (mph)	Pick-up and delivery					Line-haul					All				
			B	T	p	T4	p	B	T	p	T4	p	B	T	p	T4	p
LVM & LVD	TTC_B (s)	$25 \leq v < 55$	11.7	11.5	0.73	11.5	0.71	12.9	13.2	0.76	11.4	0.44	12.3	12.5	0.84	11.4	0.39
		$v \geq 55$															
	TTC_{min} (s)	$25 \leq v < 55$	5.0	5.0	0.9	4.9	0.35	5.0	5.3	0.28	5.4	0.33	5	5.2	0.28	5.2	0.38
		$v \geq 55$															
	Ax_{min} (m/s^2)	$25 \leq v < 55$	-1.5	-1.4	0.2	-1.5	0.55	-1.73	-1.54	0.12	-1.6	0.34	-1.62	-1.49	0.06	-1.5	0.28
		$v \geq 55$															
Lane Change	Ay_{max} (m/s^2)	$25 \leq v < 55$	1.9	2.0	0.84	1.9	0.69						1.9	1.9	0.95	1.8	0.41
		$v \geq 55$						1.4	1.4	0.58	1.3	**0.05**	1.4	1.4	0.89		
Road Departure	Ay_{max} (m/s^2)	$25 \leq v < 55$	1.9	1.8	0.4	1.9	0.72	1.7	1.7	0.80	1.7	0.69	1.8	1.7	0.79	1.8	0.96
		$v \geq 55$	1.6	1.9	0.25	1.5	0.74	1.6	2.2	0.36	2.7	0.39	1.6	2.1	0.28	2.4	0.4
	dLB_{max} (m)	$25 \leq v < 55$	0.7	0.8	0.22	0.8	0.37	0.6	0.7	0.12	0.8	0.11	0.6	0.8	**0.0**	0.8	0.06
		$v \geq 55$	0.9	0.8	0.46	0.7	0.33	0.4	0.4	0.26	0.5	0.16	0.6	0.6	0.99	0.6	0.87
	tLB (s)	$25 \leq v < 55$	2.6	2.6	0.9	2.6	0.83	2.6	2.7	0.27	2.6	0.95	2.6	2.7	0.36	2.6	0.94
		$v \geq 55$	2.2	2.5	0.27	2.4	0.44	2.4	2.4	0.91	2.5	0.42	2.3	2.4	0.28	2.5	0.24

2.4 Near-Crash Experiences

The analysis of near-crashes addresses driving conflicts that resulted in a driver response above a certain intensity level. Thus, near-crashes constitute a subset of longitudinal and lateral driving conflicts in which an intense driver response was observed during the field test data based on measures of TTC_{min}, Ax_{min}, Ay_{max}, dLB_{max}, and tLB. Near-crash thresholds were determined using distributions of intensity measures recorded in the field test. As in the previous sections, the frequency of near-crashes and driver attention behavior leading up to near-crashes are examined between the baseline and treatment test conditions by route type speed range.

The number of near-crashes is determined by applying the following thresholds to driving conflicts recorded:
- Rear-end driving conflicts – LVM, LVD and LVS focusing on braking response:
  - TTCmin less than 3 seconds and Ax_{min} over 1.962 m/s2 (0.2g) and brake pedal press over 0.5 sec, OR
  - Ax_{min} greater than 3.924 m/s2 (0.4g) and brake pedal press over 0.5 second.
- Lane-change conflict:
  - Ax_{max} greater than 0.981 m/s2 (0.1g) and 0 less than dLBmax less than 0.9 m
- Road-departure conflict:
  - Ax_{max} greater than 0.981 m/s2 (0.1g) and 0.3 below dLBmax less than 0.9 m and 1 less than tLB less than 5 sec, OR
  - Ax_{max} above 2.943 m/s2 (0.3g) and 1 less than tLB less than 5 seconds.

Analysis was conducted on driver exposure to lead vehicle moving, lead vehicle decelerating, lead vehicle stopped, lane-change, road-departure, and all other near-crashes. Two measures used for this analysis were the number of near-crashes per 1,000 miles traveled and proportion of near-crashes in driving conflicts.

All near-crashes were analyzed and coded whether or not a valid threat was present, a system alert was issued (and it was judged to be helpful in preventing a crash), or if the driver was involved in secondary tasks.

By applying the near-crash criteria presented above, the query of the processed numerical database yielded 2,472 potential near-crashes. About 26 percent of these cases did not have available video. The remaining 1,837 cases were analyzed to determine whether or not a valid threat was actually present. Of these cases, 1,672 or about 91 percent had a valid threat. The integrated system issued an alert in 860 (about 51%) of the valid near-crash cases in both baseline and treatment test conditions.

Valid near-crash cases were viewed to identify whether or not drivers were involved in secondary tasks. The proportion of valid near-crash events with secondary tasks was determined for every driver in the baseline and treatment test conditions. For pick-up and delivery drivers, the average value of this measure remained constant at 40 percent in both test conditions. In contrast, the average value increased slightly from 67 percent in baseline to 71 percent in treatment for line-haul drivers. However, this observed difference in line-haul drivers was not statistically significant based on a paired t-test ($p = 0.66$). For all drivers, the average value for the proportion of valid near-crash events with secondary tasks slightly increased from 55 percent in baseline to 57 percent in treatment with no statistical significance ($p = 0.7$).

Driver involvement in valid near-crashes was analyzed using the exposure measure of the number of near-crash encounters per 1,000 miles traveled. This analysis assumed that 91 percent of the 635 near-crashes without video had valid threats. Figure 8 illustrates the change in the number of near-crashes experienced by each driver per 1,000 miles traveled between the baseline and treatment test conditions. Positive percentages refer to a decrease in exposure from baseline to treatment, while negative percentages indicate an increase. Six pick-up and delivery drivers appeared to benefit from the use of the integrated system, since they had fewer near-crashes during the treatment period; similarly, 6 line-haul drivers also had fewer near-crashes in treatment. Overall, 12 drivers or two-thirds of all drivers experienced a reduction in near-crashes when driving with the system enabled. For all drivers, the average number of valid near-crashes per 1,000 miles decreased from 8.7 in baseline to 8.1 in treatment. This 7-percent decrease in near-crash encounter was not statistically significant ($p = 0.39$) based on the paired t-test. Pick-up and delivery drivers experienced an 8 percent decrease from 16.1 near-crashes per 1,000 miles in baseline to 14.8 in treatment ($p = 0.37$). In contrast, line-haul drivers had a smaller drop of only three percent, ranging from 2.8 to 2.7 near-crashes per 1,000 miles in treatment ($p = 0.91$). By applying the nonparametric statistical sign test to all 18 drivers, the treatment condition led to lower near-crash rates than the baseline condition, with 88-percent confidence level ($p = 0.12$). The sign test uses only the sign or direction of differences between pairs of observations in the paired-sample case, and does not take into consideration the magnitude of these differences.

It should be noted that three of the four line-haul drivers who experienced more near-crashes in treatment than in baseline had higher rates of eyes closed greater than one second, as mentioned in Section 2.2.5.1. As seen in Figure 8, line-haul drivers 23 and 24 had the most dramatic increase in near-crash encounters during the treatment condition. Assessment of the system's safety impact did not focus on the analysis of drowsy drivers, so no direct connection can be made between the role that drowsiness played in the occurrence of these near-crashes, and no definitive conclusions can be drawn about the integrated system's overall effect on levels of drowsiness. None of the 3 drivers reported that they relied on the system or that they noticed changes in their driving behavior due to driving with the system; however, driver 21 mentioned that he found the system to be most helpful in maintaining lane position when he was drowsy.

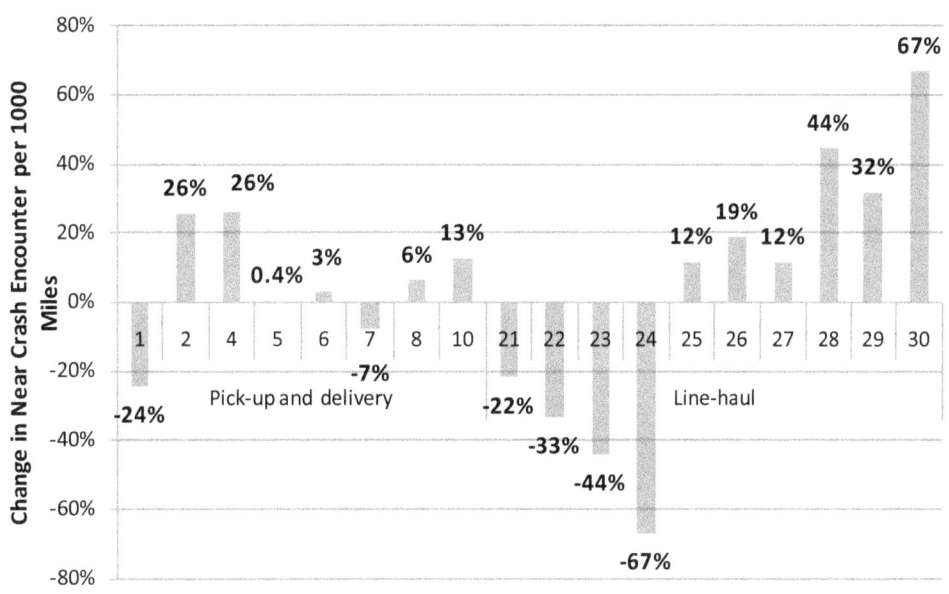

Figure 8. Change in number of near-crashes encounters per 1,000 miles in baseline versus treatment

Driver involvement in specific near-crashes was also analyzed. Figure 9 illustrates the breakdown of near-crashes by specific events and threat validity. Of cases with available video, a crash threat was present in about 99 percent of rear-end, 63 percent of lane-change, and 91 percent of road-departure near-crashes. It is interesting to note that the lead vehicle stopped scenario accounted for nine percent of all rear-end near-crashes or one near-crash per 10,000 vehicle miles traveled. In contrast, the frequency of the lead vehicle moving and lead vehicle decelerating scenarios accounted for 91 percent of all rear-end near-crashes, or 12 near-crashes per 10,000 vehicle miles traveled.

Figure 10 shows the change in near-crashes experienced by each driver per 1,000 miles traveled between test conditions. Drivers that did not experience any near-crash events in either the baseline or treatment condition were excluded from the analysis. More drivers experienced lower rates of rear-end and road-departure near-crashes in treatment than in baseline. Specifically, 8 out of 13 drivers (62%) experienced lower rear-end near-crash rates during treatment, and 11 out of 16 drivers (69%) experienced lower road-departure near-crash rates during treatment. In contrast, 8 out of 14 drivers (57%) had a higher lane-change near-crash rate in treatment than in baseline.

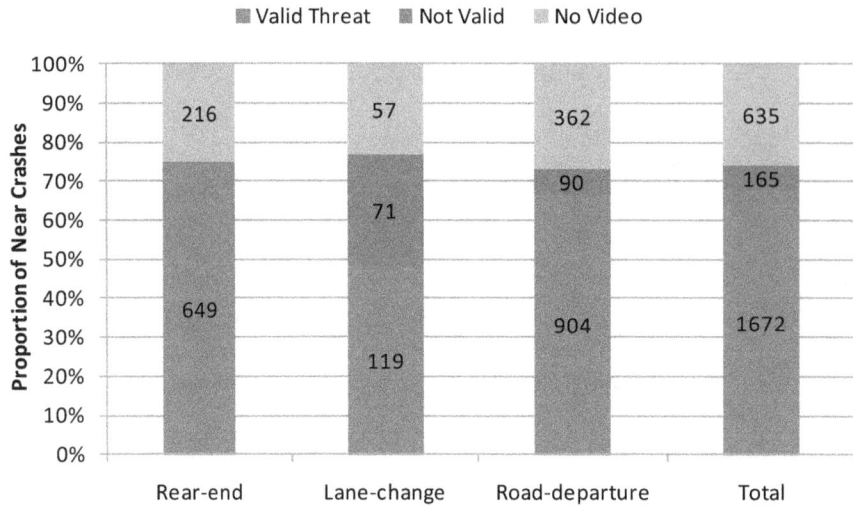

Figure 9. Distribution of near-crashes by type and threat validity

As shown in Figure 10, line-haul driver 24 showed the greatest increase in road-departure near-crashes of all drivers, while line-haul driver 23 experienced a dramatic increase in both lane-change and road-departure near-crashes between the baseline and treatment test conditions. In this figure, a negative percentage indicates an increase in near-crashes. The third line-haul driver with observed higher rates of eyes-off-forward-scene (driver 21) had the largest increase in the number of rear-end near-crashes, as well as an increase in all three types of near-crashes.

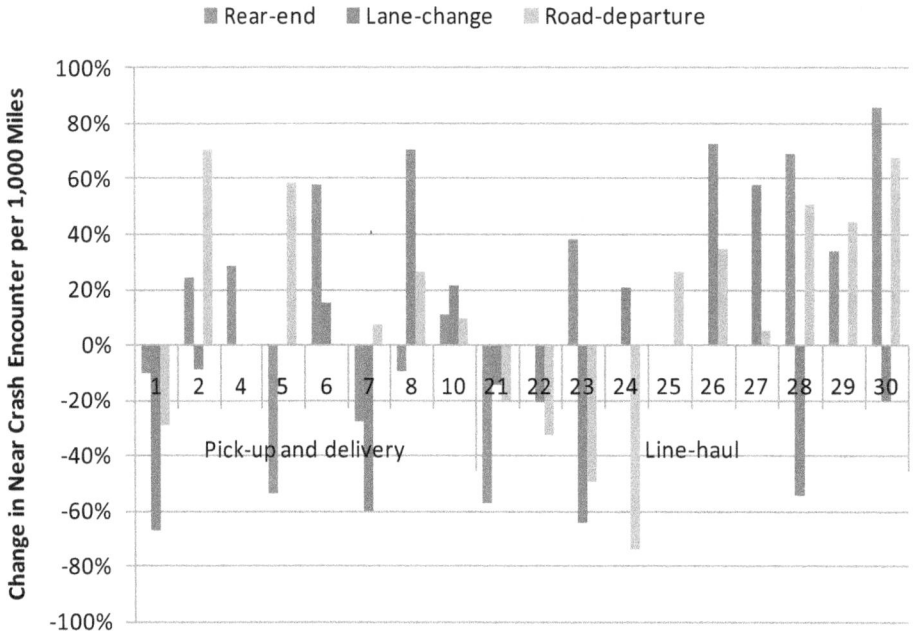

Figure 10. Change in specific near-crash rates between baseline and treatment

Table 19 presents the results of two-tail paired t-tests that were performed to observe any statistically-significant difference in the mean values of the three different near-crash rates between the baseline and treatment conditions. In addition to the mean values, the table also shows the number of observations and the p value. The 9-percent decrease in rear-end, 10-percent increase in lane-change, and 11-percent decrease in road-departure near-crash rates by all drivers were not statistically significant. By applying the nonparametric statistical sign test to all drivers, the treatment condition led to fewer road-departure near-crash rates than the baseline condition with about a 90 percent confidence level. The confidence level in the directional change of rear-end and lane-change near-crash rates by all drivers between the baseline and treatment test conditions was lower than 75 percent.

Table 19. Paired t-test results of average number of specific near-crashes per 1,000 miles driven in baseline versus treatment

Route Type	Rear-End				Lane-Change				Road-Departure			
	B	T	N	p	B	T	N	p	B	T	N	p
Pick-up and delivery	11.4	10.5	8	0.51	1.1	1.2	6	0.76	5.2	4.4	6	0.36
Line-haul	0.43	0.25	5	0.13	0.20	0.23	8	0.64	2.4	2.3	10	0.89
All	7.2	6.6	13	0.44	0.59	0.66	14	0.67	3.5	3.1	16	0.47

By excluding drivers 21, 23, and 24 (who appeared to show signs of drowsiness in a random number of video episodes), all drivers experienced a drop of 25 percent in the number of road-departure near-crashes with 92 percent confidence level from 3.7 in baseline to 2.8 in treatment. Based on the statistics of 13 test subjects who were included in this analysis, the effectiveness of the integrated safety system to potentially reduce road-departure near-crashes ranged between 6 and 46 percent.

Further analysis was conducted on road-departure near-crashes due to the relatively high confidence that the integrated safety system might have reduced this near-crash rate. These near-crashes were grouped by left and right road departures. Table 20 presents the results of the two-tail paired t-tests performed on these scenarios for pick-up and delivery, line-haul, and all drivers. It is noteworthy that the difference in the mean values of the left road-departure near-crash rate is statistically significant at the 96 percent confidence level for all drivers, and at the 91 percent level for pick-up and delivery drivers. All drivers experienced a 42 percent decrease in the near-crash rate associated with the left road-departure scenario; while pick-up and delivery drivers had a 38 percent decrease from the baseline to treatment condition. There was no statistically significant difference in the mean near-crash rates associated with the right road-departure scenario between the baseline and treatment conditions for all driver types.

Table 20. Paired *t*-test results of average number of road-departure near-crash types per 1,000 miles driven in baseline versus treatment

Route Type	Left Road-Departure				Right Road-Departure			
	B	T	N	p	B	T	N	p
Pick-up and delivery	3.3	2.1	6	0.09	2.3	2.6	5	0.76
Line-haul	0.60	0.21	6	0.35	2.0	2.2	10	0.81
All	2.0	1.1	12	**0.04**	2.1	2.3	15	0.69

Based on the analysis of 4,595 videos associated with LDW-C and LDW-I alerts issued for left lateral drifts, the host truck was drifting into an adjacent lane with opposite direction traffic in only 13 percent of the cases. In 20 percent of these opposite direction lane departure cases, another vehicle was observed approaching the host truck from the opposite direction. The time that it would take for the two vehicles to meet from the onset of the LDW alert was also determined for those cases when the opposite direction lane was occupied. This time was about three seconds or less (estimated overall response time required to avoid a collision including system warning delay, average driver response time, and vehicle response) in about 30 percent of these cases. Thus, an assumption could be made that a left lateral drift warning may have the potential to prevent an opposite direction crash in 70 percent of the cases when a vehicle drifts into an occupied lane with opposite direction traffic.

2.5 Projection of Potential Safety Benefits

This analysis projects the potential safety benefits of the integrated safety system in terms of the annual frequency of target crashes that might be avoided with full deployment of the system, N_a, where:

$$N_a = \sum_{i=1}^{n} N_{wo}(S_i) \times E(S_i) \quad (1)$$

$n \equiv$ Number of applicable pre-crash scenarios, S_i

$N_{wo}(S_i) \equiv$ Annual number of target crashes preceded by S_i prior to system deployment

$E(S_i) \equiv$ System effectiveness in avoiding target crashes preceded by S_i

Values of $N_{wo}(S_i)$ are obtained from the GES as listed in Table 1. $E(S_i)$ is expressed as:

$$E(S_i) = 1 - \frac{P_w(C|S_i)}{P_{wo}(C|S_i)} \times \frac{P_w(S_i)}{P_{wo}(S_i)} \quad (2)$$

$P_w(C|S_i) \equiv$ Probability of a crash in treatment given an S_i encounter

$P_{wo}(C|S_i) \equiv$ Probability of a crash in baseline given an S_i encounter

$P_w(S_i) \equiv$ Probability of an S_i encounter in treatment

$P_{wo}(S_i) \equiv$ Probability of an S_i encounter in baseline

The ratios $\dfrac{P_w(C|S_i)}{P_{wo}(C|S_i)}$ and $\dfrac{P_w(S_i)}{P_{wo}(S_i)}$ are known respectively as the crash prevention ratio (PR) and scenario exposure ratio (ER). The exposure ratio is obtained from the driving conflict rates observed during baseline driving and the treatment conditions. These driving conflicts map to the target pre-crash scenarios and were extracted using data mining algorithms. The prevention ratio was derived from driver performance during the baseline and treatment conditions using measures of driver response to driving conflicts. The near-crash to driving conflict proportion can also serve as a surrogate measure for computing the prevention ratio. Equations (**1**) and (**2**) can be applied to project the potential safety benefits of the integrated safety system only if statistically significant changes in driver encounter and response to driving conflicts were observed between the baseline and treatment conditions.

Analysis of indicators for unintended consequences from integrated system use revealed some statistically-significant differences between the baseline and treatment conditions at over 95 percent confidence levels. There were indications of a positive impact on safety as manifested by the increased use of turn signals by line-haul drivers during lane-change maneuvers at speeds above 45 mph, and by fewer lane excursions on both sides of the travel lane by all drivers at speeds above 35 mph. On the other hand, there were potential signs of a negative impact on safety due to a slight increase in travel speeds (about %) and more involvement in secondary tasks by line-haul drivers. However, these changes associated with line-haul drivers were offset by a 24 percent decrease in the rate of lane excursions on both sides of the travel lane at speeds above 55 mph (p = 0.03).

The experience of near-crashes in the baseline and treatment test conditions provides a good measure to estimate the potential safety benefits because it captures the frequency and severity of driving conflicts encountered during the field test. Thus, near-crash rates serve as surrogate measures for the crash prevention and scenario exposure ratios presented above. The only statistically-significant difference observed in the near-crash data was the 42-percent drop in the left road-departure scenario by all drivers from the baseline to the treatment test conditions. About 19 percent of all road departure crashes by heavy trucks were reported to occur on the left side of the road (Najm et al., 2007). As discussed above in the previous section, about 70 percent of opposite direction crashes could benefit from an integrated safety system. Based on crash statistics listed in Table 1, as well as the near-crash experience in the field test, an integrated safety system has the potential to prevent 42 percent (about 3,000) road-edge departure or no–maneuver crashes annually. In addition it could also prevent 6,000 opposite direction or no maneuver police-reported crashes annually. Given full deployment of the integrated safety system in the U.S. heavy-truck fleet, this translates to an annual reduction of approximately 4,000 police-reported crashes in these two crash types.

Considering statistically-significant differences at 85 percent confidence levels or higher, line-haul drivers had a 42 percent reduction in rear-end near-crash rates between the treatment than in baseline periods ($p = 0.13$). Applying this reduction to the annual number of about 17,000 police-reported rear-end crashes that involved a striking heavy truck at speed limits of 55 mph or higher (see Table 2), the integrated safety system could potentially prevent up to 7,000 of these crashes annually.

The safety benefits cannot be estimated for right road-departure and lane-change crashes due to the lack of statistically-significant differences in the mean values of exposure with driving conflicts and near-crashes recorded during the field test.

3. Driver Acceptance

The second goal of the independent evaluation deals with driver acceptance, which is assessed using the following five objectives:

- Ease of use: determine the usability of the integrated safety system;
- Perceived usefulness: analyze drivers' subjective assessments of safety using the integrated safety system;
- Ease of learning: assess how well drivers understand the system;
- Advocacy: determine the drivers' expressed willingness to drive a truck equipped with the integrated safety system; and
- Driving performance: monitor whether system use leads to unintended consequences, as well as any behavioral adaptations.

> **HIGHLIGHTS**
>
> - Fifteen drivers would prefer to drive a truck with the system over a non-equipped standard truck.
> - Thirteen drivers felt that driving with the system would increase their safety.
> - Fifteen drivers reported that the system made them more aware of their surroundings.
> - Fifteen drivers recommended that their employer purchase the integrated system for their fleet.
>
> Drivers found the system simple to learn and use, and auditory alerts were easy to understand.

This section presents results from the driver-acceptance analysis based on survey data. It includes the results of driver acceptance broken down by demographic and system performance variables.

3.1 Driver Acceptance Technical Approach

Driver acceptance was assessed by using subjective data in the form of survey responses. The data was quantified overall and by route type (pick-up and delivery and line-haul). Additionally, the data was separated by independent variables related to drivers' demographic information and experience with the integrated system. This section discusses the measures used to define acceptance, as well as the independent variables and methodology used in these analyses.

3.1.1 Acceptance by Driver and Objective

The five objectives of driver acceptance were rated subjectively by each test participant. Raw subjective data consist of numerical and written survey responses, verbatim comments, and results of the debriefing interview.

Most items on the post-drive survey asked drivers to rate various items on a 7-point scale with anchored points ranging from strongly disagree to strongly agree. Numerical ratings of one through three indicate a negative response, while numerical ratings of 5 through 7 indicate a positive response. A rating of four indicates a neutral response. Because the interpretation of the scale is somewhat dependent on the participant, this report provides all driver responses as positive, neutral, or negative, rather than through numerical values. The meaning of a rating six, for example, may vary from driver to driver; however, overall, ratings of 5 and greater indicate positive feelings, values below four indicate negative feelings, and a response of four is considered neutral. Quantifying survey data in this manner removes some of the individual's subjective scaling of the data.

Each survey item is mapped to a driver acceptance objective. For each driver, responses for a given objective are combined for an overall percent positive, negative, or neutral response. Results for an objective across drivers are reported as proportions of positive, neutral, and negative responses. Select survey items are also examined independently across drivers where interesting results emerged. Open-ended survey responses and verbatim comments are quantified in terms of frequency of responses across drivers.

3.1.2 Acceptance by Independent Variables

Demographic and driving history data are used to determine if any driver characteristics affected driver acceptance. Driver acceptance data are also assessed according to drivers' actual experiences with the integrated system to provide insight into whether or not the type and frequency of alerts received by drivers influenced their perception of the system.

3.1.2.1 Demographic and Driving History Variables

Demographic and driving history includes characteristics of the driver and their driving patterns. This information was obtained through a pre-drive survey that collected driver demographic information. Each driver completed this survey at the beginning of their participation in the field test.

Because of the number of drivers participating in this study and the homogeneity of the subjects, subdividing independent variables into multiple categories was not feasible. As mentioned earlier, there was not a wide range of ages and experience; most drivers were middle-aged, experienced drivers. To compensate for the lack of diversity in the subject group, the 18 drivers are grouped into two categories for each demographic variable: those with higher values; and those with lower values. The five variables and their respective groupings are shown below.

Table 21. Driver demographic categories used in driver-acceptance analysis

Demographic	Group	Number of Drivers
Route type	Pick-up and delivery	8
	Line-haul	10
Age	Greater than 49	9
	Less than 49	9
Years with CDL	Greater than 25	9
	Less than 25	9
Traffic violations	Yes	8
	No	10
Prior experience with advanced safety systems	Yes	6
	No	12

3.1.2.2 Driver Experience Variables

The variables of the driver experience represent metrics about the types of alerts the drivers received while driving with the integrated system, alert characteristics, and the temporal distribution of the alerts. All experience metrics refer to the system performance in the treatment period only since the performance during this time period is the sole basis of drivers' subjective responses. Unless otherwise noted, all alert rates refer to the number of alerts per 100 miles.

Each variable is broken down into higher and lower groups as discussed in the previous section; however, rather than dividing the drivers into even groups, the mean value within the group was used as a threshold to create groups. In circumstances where the mean produced an uneven grouping (a group less than 8 and more than 10), the median value was used to reduce the size difference between the groups. Table 22 lists each driver experience variable and provides the number of drivers in each group, and the threshold used to group the drivers.

Table 22. Driver experience categories used in driver-acceptance analysis

Experience Metric	Group	Number of Drivers	High/Low Threshold
Overall Alert Rate	Low	10	Mean
	High	8	(19.8)
FCW Rate	Low	9	Median
	High	9	(3.6)
LCM Rate	Low	10	Mean
	High	8	(2.4)
LDW-I Rate	Low	9	Mean
	High	9	(7.6)
LDW-C Rate	Low	10	Median
	High	8	(3.6)
Overall Incorrect Targets	Low	9	Mean
	High	9	(37%)
% FCW with No in-Path Target	Low	9	Median
	High	9	(55%)
% LCM with No Adjacent Vehicle	Low	9	Mean
	High	9	(47%)
% LDW-I with No Adjacent Target	Low	9	Mean
	High	9	(44%)
% LDW-C with No Lane Excursion	Low	9	Median
	High	9	(9%)
Overall Conflict Rate	Low	10	Mean
	High	8	(8.7)
Rear-End Conflict Rate	Low	10	Mean
	High	8	(6.7)
LCM Conflict Rate	Low	9	Median
	High	9	(0.4)
Road Departure Conflict Rate	Low	9	Median
	High	9	(1.2)
% Alerts with Conflicts	Low	10	Mean
	High	8	(5.9%)

3.1.2.3 *Driver Acceptance by Demographics and System Experience*

This analysis is conducted by examining survey responses for each independent variable in order to extract variations in driver acceptance between both groups of drivers. For each numerical survey item, the mean response of drivers within each of the two groups associated with an independent variable is determined, and effect size is calculated using the mean difference and the pooled standard deviation (effect size equals the difference between the means or pooled standard deviation). The sample size of this study is small for a between subjects comparison, so statistical significance is not tested for in this analysis. While effect size does not indicate a significant difference between the means, it does indicate trends between groups. For this analysis, a stringent effect size of 0.8 is selected to indicate a group trend.

3.2 Survey Results

General results, as well as results within each driver-acceptance objective, are presented based on survey responses. Aggregate results for numerical response type questions are provided for "ease of use" and "perceived usefulness" objectives. Other objectives with fewer related survey items are discussed in terms of the results of individual survey items. Verbatim comments obtained during the debriefing interviews are also included where appropriate.

3.2.1 General Feedback

Drivers' responses to the open-ended question "What did you like most about the integrated system?" are described in Table 23. Five drivers commented that what they liked most about the integrated system was the blind spot monitor displays that assisted them when making lane changes. An equal number of drivers also liked the forward warning system (combination of FCW alerts and headway warnings). Four drivers also reported liking the lane departure warning system best. There were differences in the responses by route type; line-haul drivers were more likely to report liking the BSM displays (4 of 10 drivers) and pick-up and delivery drivers were more likely to say that they liked the general sense of increased awareness and alertness best (3 of 8 drivers).

Table 23. System features most liked by drivers

Feature	Number of drivers	
	Pick-up and Delivery	Line-haul
Forward collision warning (FCW)	1	1
Lane-change/merge (LCM)	1	-
Lane departure warning (LDW)	1	3
Blind spot monitor (BSM)	1	4
Increased alertness/ awareness	3	-
Headway warning	1	2

Table 24 summarizes drivers' responses to the open-ended question "What did you like least about the integrated system?" Seventeen drivers said what they liked least about the system were the false warnings. The most common response was false FCW alerts (5 of 8 pick-up and delivery drivers, 2 of 10 line-haul drivers). While pick-up and delivery drivers were more likely to report disliking false FCW alerts, line-haul drivers primarily disliked false side hazard alerts (4 of 10 line-haul drivers), which include LCM and LDW-I alerts.

Table 24. System characteristics least liked by drivers

Feature	Number of drivers	
	Pick-up and Delivery	Line-haul
False forward collision warning (FCW)	5	2
False lane departure warning (LDW)	1	1
False side hazard	-	4
Headway warning too sensitive	1	-
False warnings (general)	1	2

Table 25 lists the results of the open-ended question "In which situations did you find the integrated system to be most helpful?" Four line-haul drivers found the system to be most helpful when drifting out of their lane. Pick-up and delivery drivers and line-haul drivers found the system to be helpful when approaching slowed or stopped traffic. Two pick-up and delivery drivers thought the system was most helpful when vehicles were in their blind spots.

Table 25. Number of drivers who found the system to be most helpful in driving situations

Pick-up and delivery		Line-haul	
Approaching slower/stopped traffic	2	Drifting	4
Cars in blind spots	2	Approaching slower/stopped traffic	3
Freeway driving	1	Cars in blind spots	1
Heavy traffic	1	Changing lanes	1

3.2.2 Ease of Use

Results of the "ease of use" objective are shown in Figure 11. This figure shows the percent positive, neutral, and negative responses to 12 survey items pertaining to ease-of-use. For this objective, 5 of 8 pick-up and delivery drivers had favorable opinions of the integrated system. Five of 10 line-haul drivers responded positively to ease of use questions.

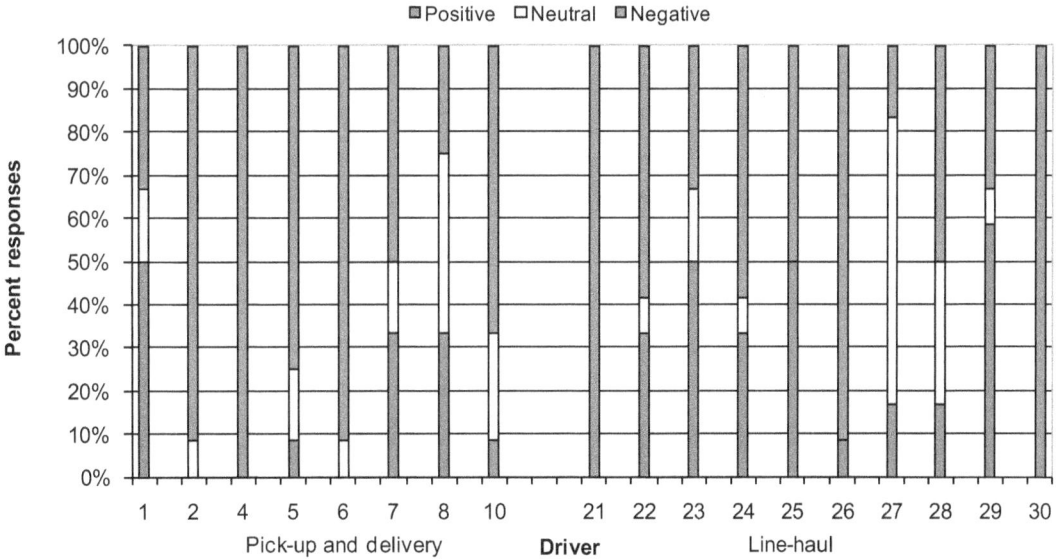

Figure 11. Aggregate results of 12 survey items related to ease of use

Figure 12 illustrates drivers' overall satisfaction with the integrated system. Over half of both pick-up and delivery and line-haul drivers were satisfied with the integrated system. Only two drivers reported being dissatisfied with the system, one of whom commented that he preferred not to have any technology in his truck in general.

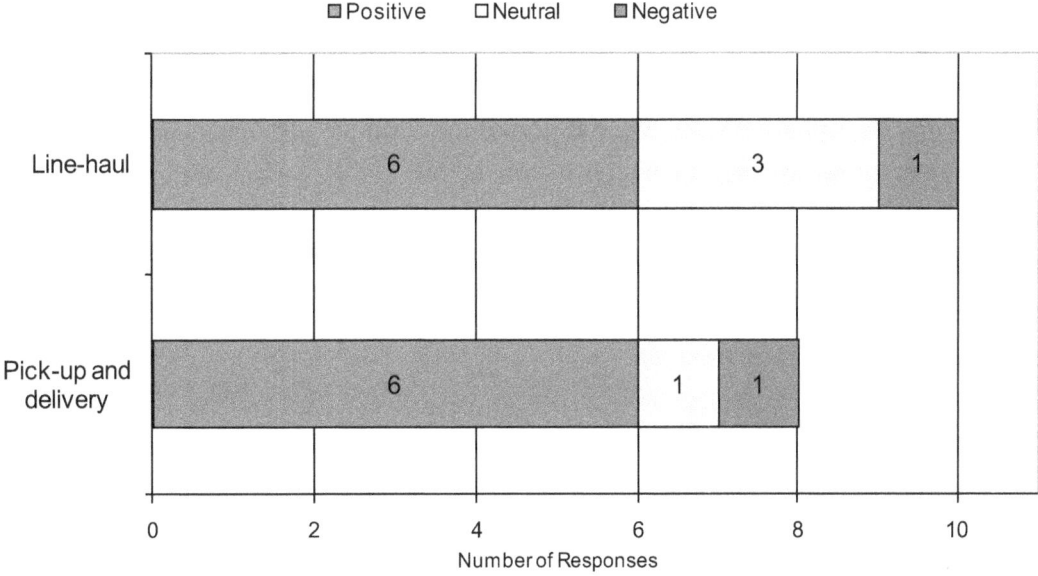

Figure 12. Responses to the question "How satisfied were you with the integrated system?"

Figure 13 shows drivers' responses to the survey item "The integrated system made my job easier." Half of the pick-up and delivery drivers and half of the line-haul drivers felt that the system made their job easier. In the debriefing interview, drivers who responded negatively to this question were asked if they felt that the system actually made their job more difficult; none agreed with that statement. One driver who responded negatively made the following comment: "It is helpful, but I am still in charge of the truck and still need to drive the same (as I do without the system)."

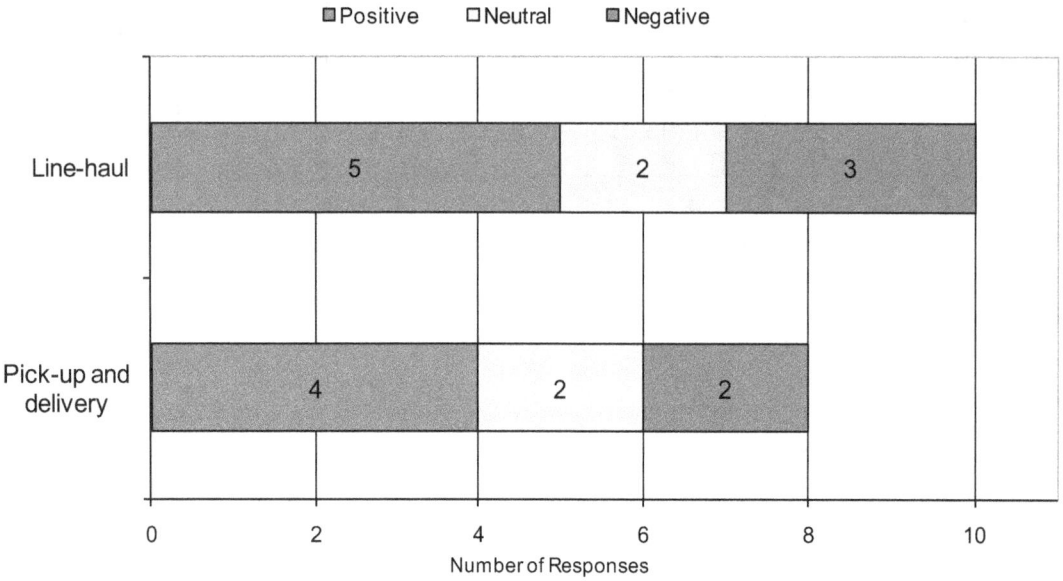

Figure 13. Responses to the statement "The integrated system made my job easier"

Figure 14 shows drivers' opinions about the usability of the auditory warnings. Seventeen drivers said that they could easily distinguish between the two warning sounds (one to indicate a forward threat, one to indicate a lateral threat). Sixteen drivers thought that the auditory warnings were attention getting. These responses did not differ significantly between pick-up and delivery and line-haul drivers. Overall, the auditory alerts were judged to be salient and effective.

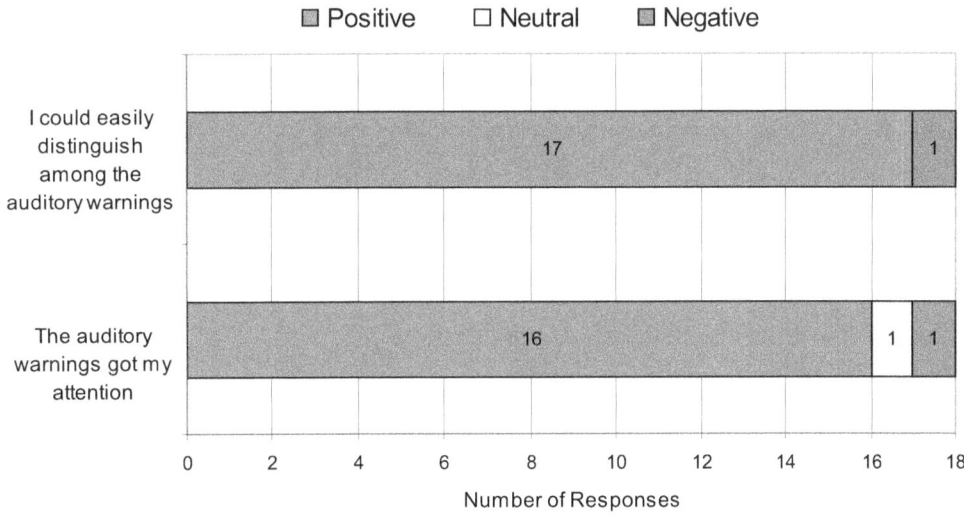

Figure 14. Drivers' opinions of the integrated system auditory warnings

Figure 15 illustrates responses to the statement "The auditory warnings were not annoying." Line-haul drivers reported more annoyance with the auditory warnings than the pick-up and delivery drivers. Many of the drivers who reported annoyance with the auditory warnings commented that the sound of the warnings was not annoying, but rather the fact that they received warnings when they did not want them. One pick-up and delivery driver noted: "I was getting them when I didn't want them, it gave me alerts for things I already saw coming." Overall, line-haul drivers received many more alerts than pick-up and delivery drivers, which could be the reason for increased annoyance with the auditory tones.

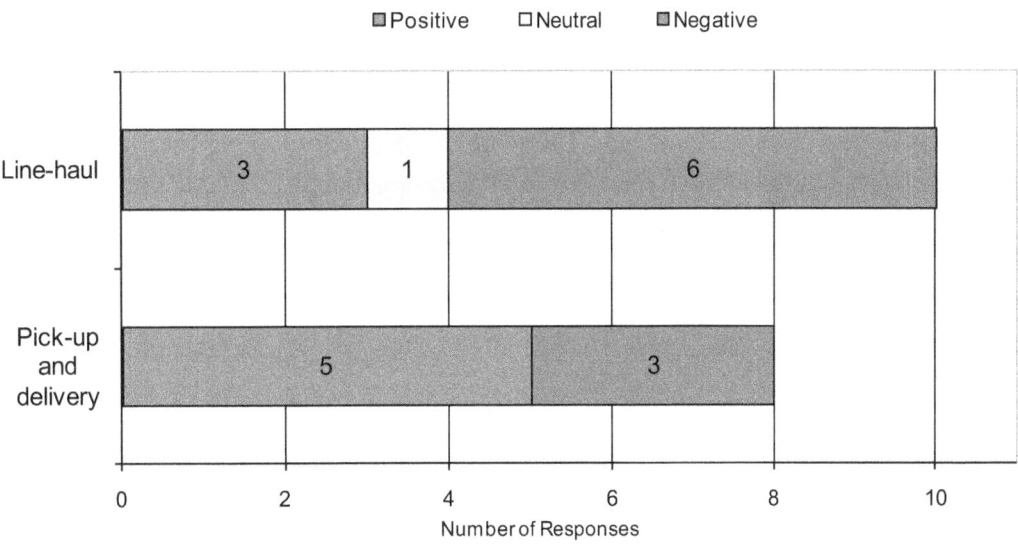

Figure 15. Drivers' responses to the survey item "The auditory alerts were not annoying"

3.2.3 Perceived Usefulness

Aggregate results of each driver's opinion of the usefulness of the integrated system are illustrated in Figure 16. These results are based on the responses of 10 Likert-scale questions associated with usefulness of the system. Six of 8 pick-up and delivery drivers felt that the system was useful overall, and 1 driver responded favorably to half of the survey items related to usefulness. Seven of the 10 line-haul drivers responded positively to the usefulness of the system overall, and one line-haul driver responded positively to half of the questions. Overall, 3 drivers responded positively to less than half of the questions.

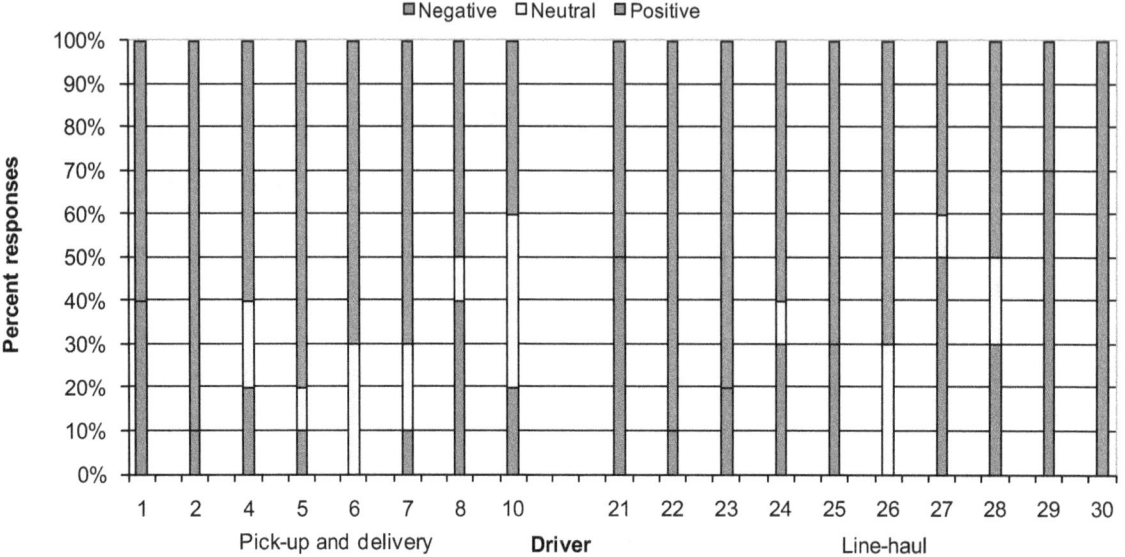

Figure 16. Aggregate results of 10 survey items related to perceived usefulness

Figure 17 shows drivers' opinions of the increase in safety due to their use of the integrated system. Most drivers felt that they received a safety benefit from driving with the integrated system, and that the system made them more aware of their surroundings and the position of their truck in the travel lane. There was no significant difference in responses between pick-up and delivery and line-haul drivers.

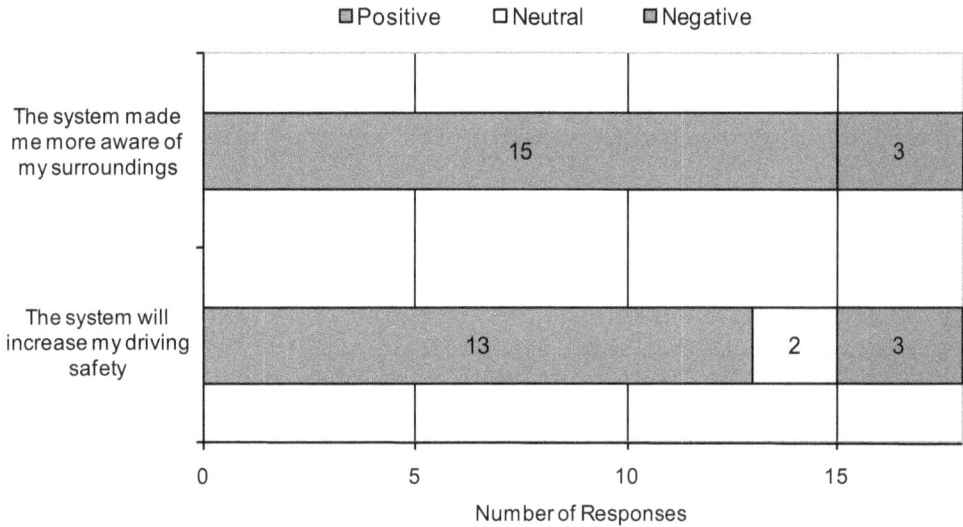

Figure 17. Drivers' opinions on the safety increase associated with the system

In an attempt to determine the effect of nuisance warnings, drivers were asked to rate their agreement with the statement: "The number of false warnings caused me to begin to ignore the system." A response of "yes" indicates low trust in the system due to receiving a large number of alerts that the driver did not find helpful. As shown in Figure 18, more line-haul drivers than pick-up and delivery drivers reported ignoring the system warnings due to the false alarms. While pick-up and delivery drivers generally drive in high traffic areas with many surrounding vehicles and many vehicle maneuvers, most of the mileage driven by line-haul drivers is on freeways at night with a very low probability of encountering an obstacle. These differences in route type could explain the variation in response between line-haul and pick-up and delivery drivers.

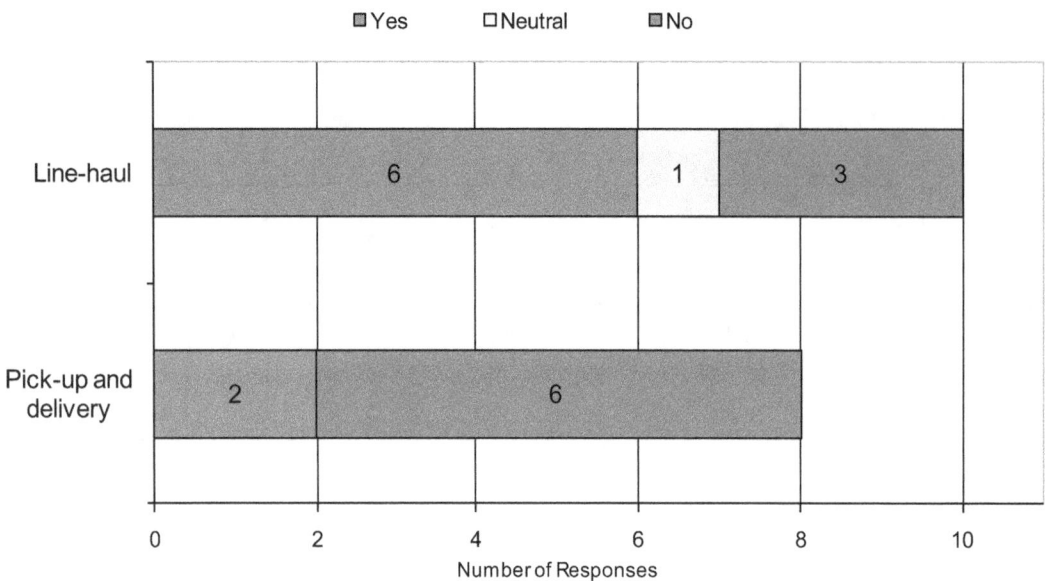

Figure 18. Responses to "The number of false warnings caused me to ignore the integrated system"

Figure 19 shows drivers' perceptions of the presence of each type of false warning. Drivers were asked the same question for the system overall, as well as for each type of warning. This method allows the comparison of responses across alert type. Both driver groups gave similar responses for the presence of false alarms overall, while pick-up and delivery drivers were more likely to report the presence of hazard ahead warnings (FCW), line-haul drivers reported more side hazard (LDW-I/LCM) and drift (LDW) warnings. These differences are likely the result of differences in roads travelled between the two route types.

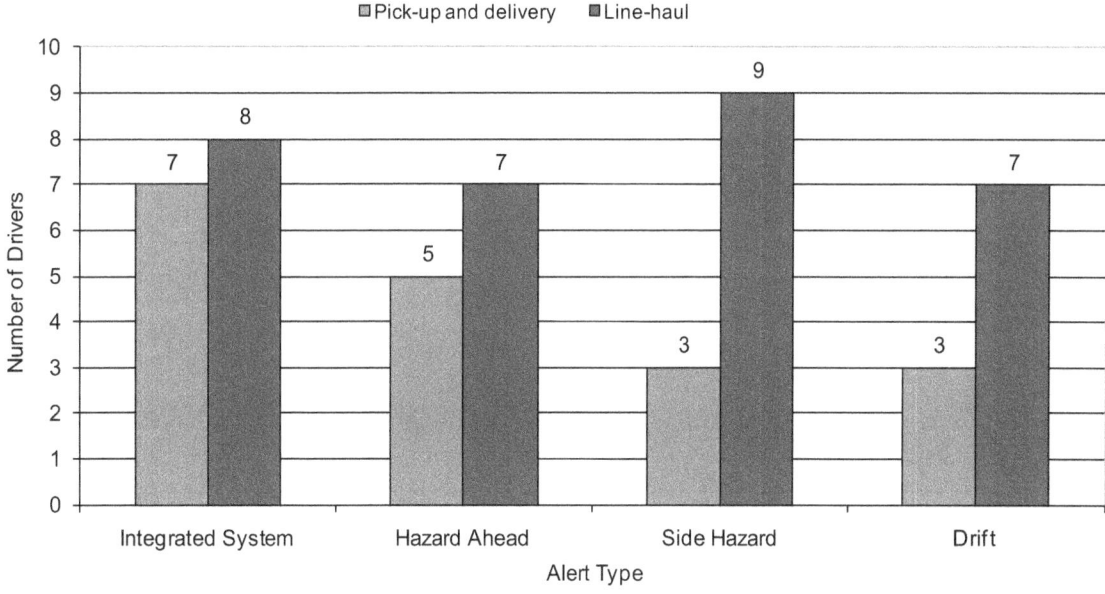

Figure 19. Responses to "The integrated system gave me alerts when I did not need them"

Eight drivers reported that the integrated system helped them from getting into a crash or near-crash. Drivers listed the following scenarios in their responses: driving while fatigued (LDW assisted in staying on the road); alerting the driver of a slowing vehicle with no brake lights; alerting drivers of vehicles in their blind spot during lane changes; and making drivers aware of a vehicles that had cut in front of them.

3.2.4 Ease of Learning

Drivers were asked three questions addressing ease of learning on the post drive survey: one numerical response, one yes-no question, and one open-ended question.

Drivers were asked to respond to "I understood what to do when the integrated system provided a warning" as a way to gauge if drivers understood the meaning of the various warning modalities. As shown in Figure 20, 16 drivers said that they understood the warnings.

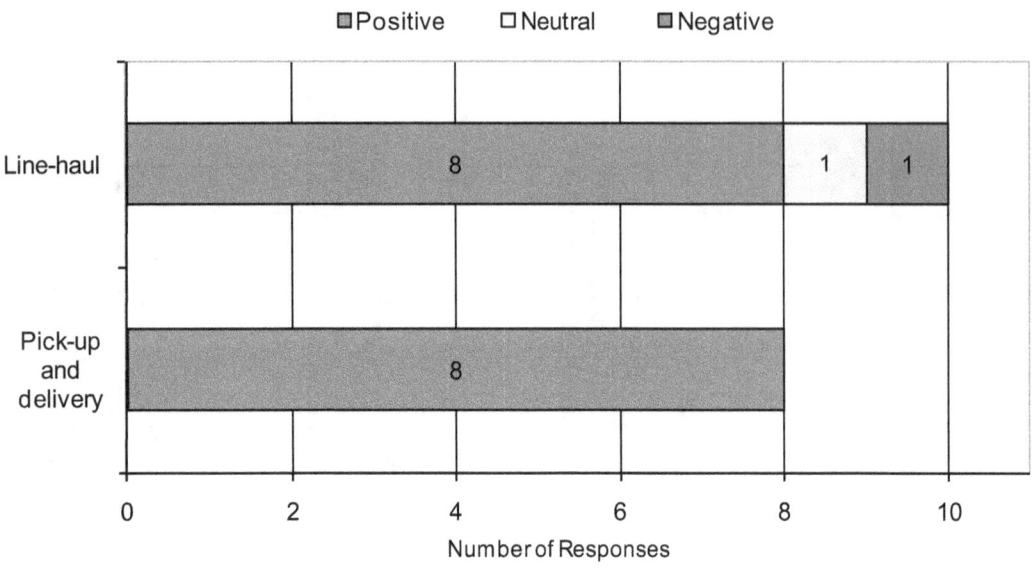

Figure 20. Responses to the question "I always knew what to do when the system provided me with a warning"

When asked if the integrated system performed as they had expected, all pick-up and delivery drivers said that it did. They all had a good understanding of what the system was supposed to do and how it worked. However, 5 out of 10 line-haul drivers said that the system did not perform as they expected it to. All 5 line-haul drivers who provided a response of "no" commented that they received more false warnings than they originally thought they would, and that caused confusion about how the system worked.

A total of 17 drivers reported getting used to having the system in their vehicle within a week. Drivers commented that the system concept was very simple and easy to learn. The remaining driver said that it took three weeks to fully understand the system and to be able to predict when the alerts would occur.

3.2.5 Advocacy

One numerical response and two yes-no questions in the post-drive survey addressed whether or not the drivers would recommend the purchase of an integrated system by their company. To gauge drivers' general attitudes towards advanced technology, they were asked to respond to this statement: "In general, I like having new technology in my truck." As shown in Figure 21, only two drivers (one line-haul and one pick-up and delivery) replied that they did not like the idea of having technology in their trucks. Both of these drivers commented that with or without technology they were responsible for the truck. One of the drivers (pick-up and delivery) referred to himself as "old school" and said that he felt that some technologies can even be dangerous (such as electronic stability control) and he would rather have complete control over the vehicle. In general, this driver's opinions of the integrated system were negative, whereas the line-haul driver that did not like technology in general had positive opinions of the system overall.

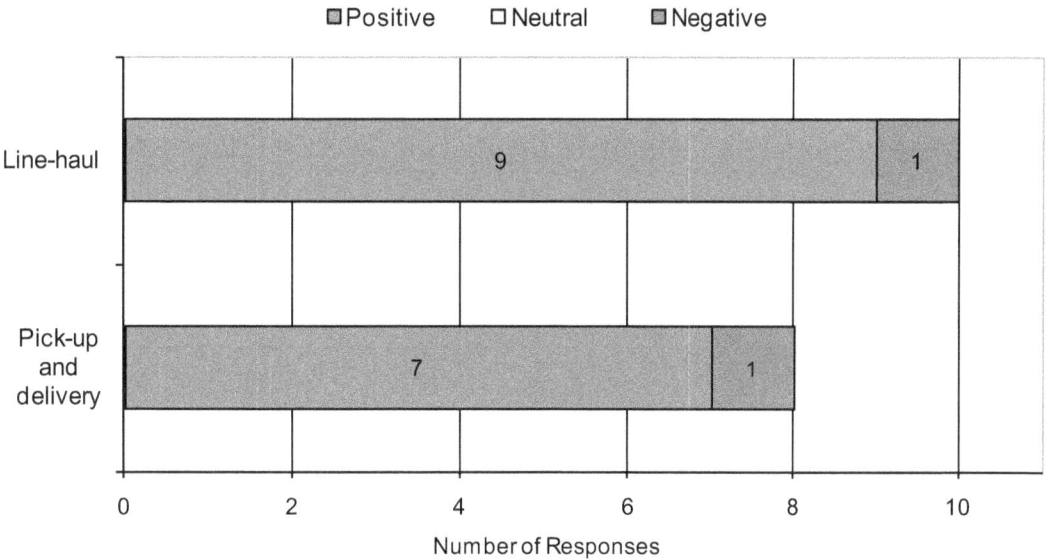

Figure 21. Drivers' opinions of having new technology in their trucks

Drivers were asked two yes-no questions pertaining to advocacy for the system including, "Do you prefer to drive a system with the integrated system over a conventional truck?" and "Would you recommend that the company buy trucks equipped with the integrated system?"

The first question aimed to understand if, overall, each driver would prefer to drive with or without the system. Fifteen drivers said they would prefer to drive with the integrated system than without. Responses of line-haul and pick-up and delivery drivers are shown in Figure 22. One of the two pick-up and delivery drivers who said he would prefer to drive without the system said that he did not feel it was helpful for pick-up and delivery drivers, but if he were a line-haul driver he would like to drive with the system. The line-haul driver who did not want to drive with the system commented that "the system made too much noise and gave too many false warnings." Overall, about half of the drivers commented that the reason they would prefer to drive with the system is that it increased their alertness and made them feel safer. One pick-up and delivery driver commented: "with all the distractions we have during an extremely hectic day, it's like having an extra set of eyes looking out for you."

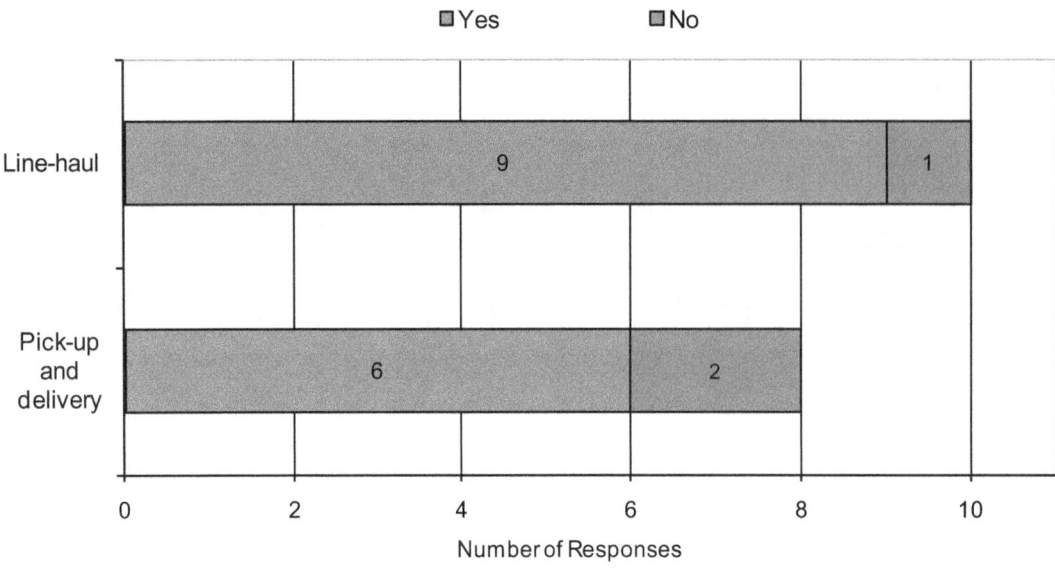

Figure 22. Drivers' willingness to drive with the integrated system

Responses to the question "Would you recommend that the company buy trucks equipped with the integrated system?" were primarily favorable, as well. Fifteen drivers would recommend the system to their employer, as shown in Figure 23. Most drivers felt that despite the system's flaws, the system provided an overall safety benefit that would be an asset to their companies. One driver said: "In the long run it would save the company money and help give us a good name as a company with safe and accident-free drivers." The 2 line-haul drivers who would not recommend the system felt that the system had too many false warnings. One commented: "I am not sure the cost is justified because of the false warnings."

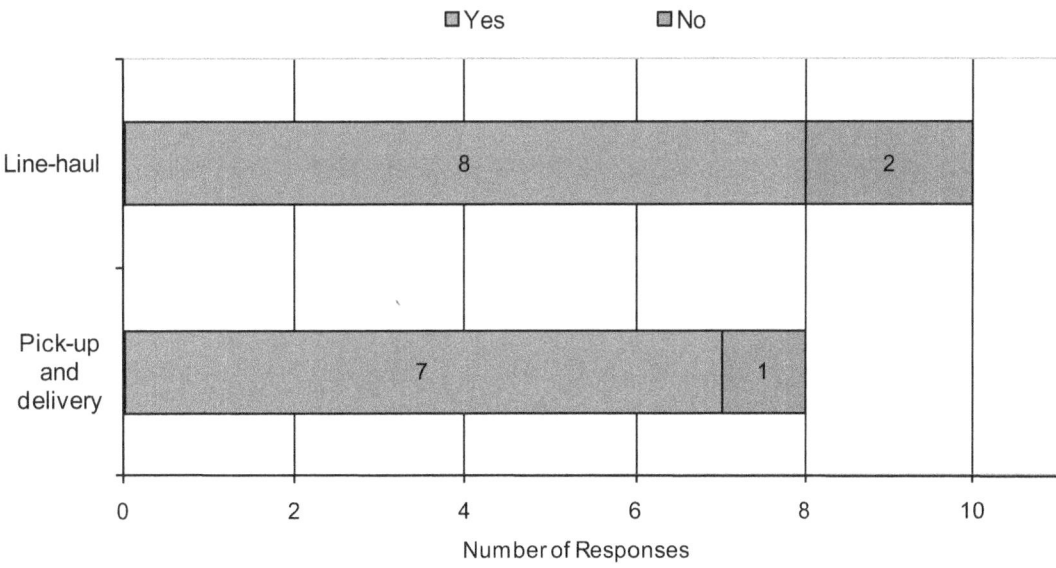

Figure 23. Driver's willingness to endorse the integrated system to their employer

3.2.6 Driving Performance

Two questions were included to assess how the integrated system affected driving performance, with the intent of soliciting feedback on whether or not driving with the system would create any unintended consequences.

The first question "Did you rely on the integrated system?" showed that most drivers were not relying on the system, as illustrated in Figure 24. Those who did report relying on the system specifically said that they relied on the system to help them stay alert in heavy traffic (1 pick-up and delivery driver), to help them stay in their lane (1 line-haul driver), and to help make lane changes in bad weather (1 line-haul driver).

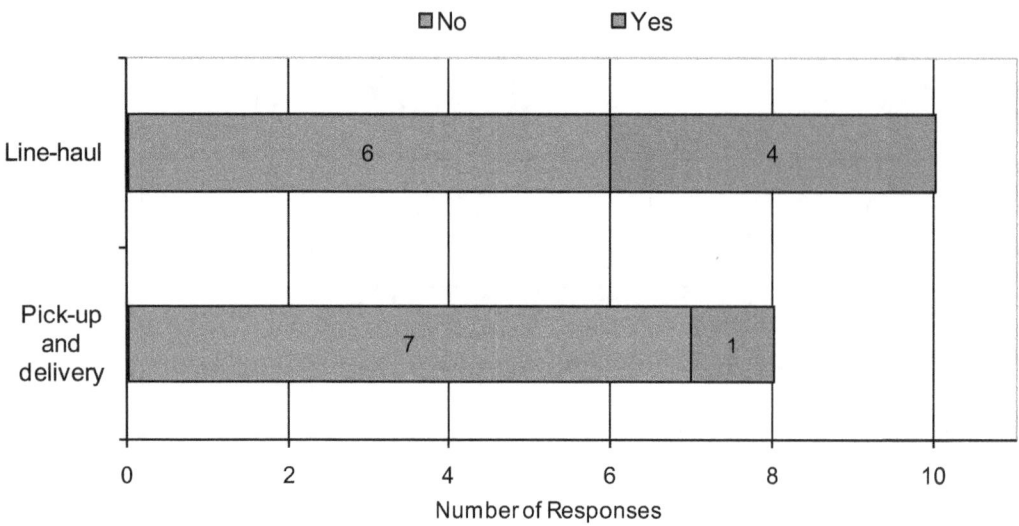

Figure 24. Drivers' self-reported reliance on the integrated system

Table 26 shows drivers' self-reported changes in driving behavior due to the integrated system. 8 of 10 line-haul drivers reported no changes in their driving behavior, and three of 8 pick-up and delivery drivers reported more changes. The changes in driving behavior reported by the drivers were all positive and including increased alertness, increased following distance, and an improvement in lane positioning. Two pick-up and delivery drivers also commented that they increased their turn signal use.

Table 26. Changes in driving behavior due to integrated system use

Driving Behavior	Number of drivers	
	Pick-up and Delivery	Line-haul
Increased alertness/ awareness	3	1
Increased headway	2	-
Improved lane positioning	-	1
None	3	8

3.3 Driver Acceptance by Demographic Variables

This analysis explores the differences in driver opinion based on driver characteristics. The results below list survey items showing a group trend in the means by the following five independent variables:

- Route type:
  - Line-haul drivers were more likely than pick-up and delivery drivers to say the number of false warnings caused them to ignore the integrated system's warnings (effect size = 1.14).
  - Line-haul drivers agreed much more strongly than pick-up and delivery drivers that the system gave them side-hazard warnings when they did not need them (effect size = 1.18).
  - Five of 8 pick-up and delivery drivers noticed changes in their driving behavior due to the system, but only 2 of 10 line-haul drivers reported changes in their driving behavior.
  - All 8 pick-up and delivery drivers said that the system performed as they expected, but the system met expectations for only 6 of 10 line-haul drivers.

- Traffic offenses in the previous three years:
  - Drivers who had not had any traffic offenses for the previous 3 years were more likely to report that the number of false warnings they received affected their ability to correctly understand the integrated system (effect size = 1.01).
  - Drivers who had traffic violations were less distracted by the system warnings (effect size = 1.10).
  - Drivers who had traffic offenses reported more strongly that the system got their attention than drivers with no traffic offenses (effect size = 0.87).
  - Drivers with no traffic offenses found the blind spot warning lights to be more annoying than drivers with traffic offenses (effect size = 0.95).
  - Half of the drivers who had no traffic violations (5 of 10) reported changes in their driving behavior due to the system. Only 2 of the 8 drivers who had traffic violations reported that driving with the system changed their driving behavior.

- Age:
  - Younger drivers were more likely than older drivers to report receiving system warnings when they did not need them (effect size = 1.01).
  - Younger drivers were more likely than older drivers to report receiving FCW alerts when they did not need them (effect size = 0.97).
  - Most of the older drivers (7 of 9) reported no changes in their driving behavior due to the integrated system, but 5 of 9 younger drivers said that driving with the system caused changes in their driving behavior.

- Years with commercial drivers license (CDL):
  - Drivers licensed for 25 years or more were more likely to report being distracted by the integrated system warnings (effect size = 0.86).
  - Drivers licensed for 25 years or more were less likely to report changes in their driving behavior than drivers who had had their license for a shorter period of time (2 of 9 compared to 5 of 9).

- Prior experience with advanced safety systems:
  - Drivers with prior experience with advanced safety systems were more likely to report changes in their driving behavior than drivers without prior experience (1 of 6 versus 6 of 12).

3.4 Driver Acceptance by Driver Experience Variables

This analysis explores the differences in driver opinion based on driver experience in the field test. The results below list survey items showing a group trend in the means by the following four measures:

- Alert rate:
  - Drivers with lower overall alert rates reported that the number of false warnings affected their ability to correctly understand the system (effect size = 0.81).
  - Drivers with higher rates of LCM warnings were more likely to report that the number of false warnings affected their ability to correctly understand the integrated system (effect size = 0.92).
  - Drivers with lower FCW or LDW-I rates were more likely to agree that the number of false warnings caused them to begin to ignore the integrated system (effect size = 0.88 for both)
  - Six of 10 drivers with lower alert rates reported changes in their driving behavior due to driving with the integrated system, while only one of 7 drivers with higher alert rates reported changes.
  - Eight of 10 drivers with lower alert rates said that the system performed as expected. Only three of 8 drivers with higher alert rates agreed.

- Invalid alert proportion:
  - Drivers with overall higher invalid alert rates agreed more strongly that they always knew what to do when the integrated system provided them with a warning.
  - Drivers with higher overall invalid alert proportions and a higher proportion of LCM alerts felt more strongly that they knew what to do when the integrated system provided them with a warning (effect size = 1.03).
  - Drivers with lower overall, FCW, and LCM invalid alert proportions were more likely more likely to report that the number of false warnings they received caused them to ignore the integrated system (effect sizes = 1.44, 1.05, and 1.44, respectively).
  - Drivers with higher invalid alert proportions were less likely to say that they found the false warnings to be annoying (effect size = 0.99).
  - Drivers with lower invalid alert proportions felt more strongly that they always knew what to do when the system provided them with a warning (effect size = 0.87).
  - Drivers with lower invalid LCM proportions agreed more strongly that the integrated system provided them with false LCM warnings (effect size = 1.88).

- Drivers with a lower proportion of invalid drift warnings agreed more strongly that the number of false warnings affected their ability to correctly understand the integrated system (effect size = 0.85).
- All 9 drivers with lower invalid alert rates said the system performed as they expected it to, but only about half of drivers with higher alert rates (5 of 9) said that the system performed as they expected.
- Most drivers with lower invalid alert rates (7 of 9) reported no changes in their driving behavior due to the presence of the integrated system, but over half (5 of 9) drivers with lower invalid alert rates reported changes in behavior.

- Conflict rate
  - Drivers with higher conflict rates overall, and higher rates of each of the three conflict types (rear-en, lane change/merge and road departure) were less likely to say that the number of false warnings caused them to begin to ignore the integrated system (effect size = 1.14, 1.14, 1.05 and 0.88 respectively).
  - Drivers with lower road departure conflicts agreed more strongly that the number of false warnings they received made it difficult to understand the integrated system (effect size = 1.05).
  - Drivers with both higher lane change/merge conflict rates and higher road departure conflict rates were less likely to find the false warnings annoying (effect size = 0.84, 0.99)
  - Drivers with higher lane change/merge conflict rates were less likely to report that the system gave them false side hazard warnings (effect size = 1.35).
  - Drivers with higher road departure conflict rates were less likely to report that the system issued false drift warnings (effect size =1.07).
  - Drivers with both higher lane change/merge conflicts and higher road departure conflicts agreed more strongly that they always knew what to do when the integrated system issued a warning (effect size =0.87, 0.87).
  - Drivers with higher rates of road departure conflicts found the lane departure warning availability icons more useful than drivers with lower road departure conflicts (effect size=0.89).
  - Seven of 8 drivers with high conflict rates said that they did not rely on the integrated system, while 4 of 10 drivers with lower conflict rates said they did rely on the system.
  - Two of 10 drivers with lower conflict rates said that they noticed changes in their behavior due to use of the integrated system, while 5 of 8 drivers with higher conflict rates reported that they had changed their behavior due system use.
  - All 8 drivers with high conflict rates thought that the system performed as expected, but only 6 of 10 drivers with lower conflict rates agreed.

- Proportion of alerts corresponding to conflicts:
  - Drivers with a higher proportion of their alerts corresponding to conflicts (true alerts) were less likely to report that the false warnings were annoying (effect size =1.02).
  - None of the 8 drivers with higher proportion of their alerts corresponding to conflicts said that they relied on the integrated system. Half of the drivers (5 of 10) with a lower proportion of their alerts corresponding to conflicts said that they relied on the integrated system.
  - All 8 drivers with a higher proportion of their alerts corresponding to conflicts said the system performed as they expected, while only 6 of 10 drivers with lower proportions agreed.

4. System Capability

This section provides results of the system capability analysis that was conducted for the sensors, warning logic, driver-vehicle interface, and robustness of the integrated system. The performance of the forward-looking, side-looking, and lane-tracking sensors was evaluated in terms of their ability to detect targets in the path of the host heavy truck, while rejecting out-of-path targets, and determine truck position within the travel lane. The warning logic was examined in terms of the system's decision making to alert drivers to driving conflicts that might lead to rear-end, lane-change/merge, or road-departure crashes. The driver-vehicle interface was evaluated in terms of its capability to properly convey visual and audible information to the driver. System robustness is appraised by its availability during the field test.

> **HIGHLIGHTS**
> - Out-of-path targets accounted for 7 percent of FCW alerts for moving targets, while 97 percent of FCW alerts for stopped objects were due to misclassification of roadside objects and bridges.
> - Over 50 percent of side-imminent alerts were issued when no target was present in the warning zone.
> - Ninety percent of the lateral-drift cautionary alerts were issued when lane boundaries were crossed without the use of turn signals.
> - The lateral drift warning function met system specifications for all speed ranges.
> - The FCW alert rate for moving targets was 12 percent lower with the system enabled.
> - Lateral-drift cautionary alert rates were 21 and 17 percent lower for left- and right-lane excursions, respectively, with the system enabled.
> - Based on the reduction in FCW-M and LDW-C alert rates, the system could potentially prevent 13,000 heavy-truck crashes annually.

4.1 Sensors

This analysis is based on a sample of 12,900 alert videos which were reviewed to characterize the performance of the integrated system's sensor suite. A detailed breakdown of the alerts analyzed is located in Appendix D, and definitions of each coded variable discussed in this section are located in Appendix E.

4.1.1 Forward-Looking Sensors

Evaluation of forward-looking sensor performance was based on the analysis of 2,368 FCW alerts characterized by the location of detected objects (targets). Target location refers to whether the detected object was in the equipped vehicle's intended lane of travel at the time of the alert, or just prior to the alert being issued. The system was designed to issue alerts for in-path objects only. Roadside signs, overhead bridges, guard rails, and vehicles in adjacent lanes were all considered out-of-path targets. System performance was measured by the proportion of FCW alerts (FCW-M and FCW-S) issued for out-of-path targets. The distribution of alerts issued for out-of-path targets listed by target type, host truck maneuver, position, and location (as defined in the video coding manual of the MDAT in Appendix D) is provided later in this section.

As shown in Figure 25, about 97 percent of FCW-S alerts were issued for out-of-path targets. In contrast, only 7 percent of all FCW-M alerts were issued for objects that were not in the equipped vehicle's intended lane of travel. Figure 26 illustrates the distribution of FCW-S alerts by out-of-path target type for all drivers. For pick-up and delivery drivers, FCW-S alerts were issued most frequently for roadside signs (64%), while FCW-S alerts for overhead bridges were often issued for line-haul drivers (44%). Pick-up and delivery drivers received about 43 percent of FCW-S alerts for out-of-path targets when negotiating a curve. On the other hand, line-haul drivers received about 92 percent of these alerts when traveling on straight roads. For curve-related FCW-S out-of-path alerts, 55 percent occurred while all drivers were within the curve, as opposed to about 36 percent at curve entry, and 9 percent at curve exit. Construction zones accounted for 12 percent of all out-of-path FCW-S alerts, and ramps made up 4 percent of all FCW-S out-of-path alerts.

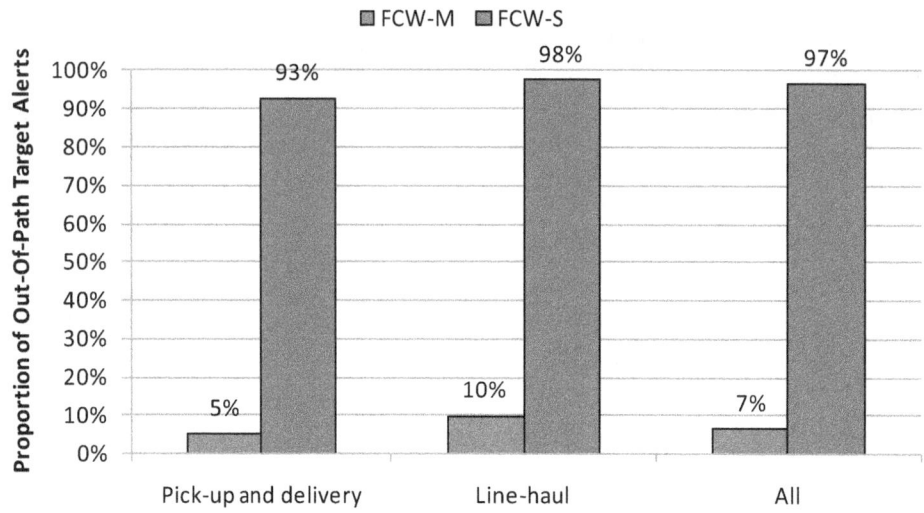

Figure 25. Proportion of FCW alerts triggered by out-of-path targets

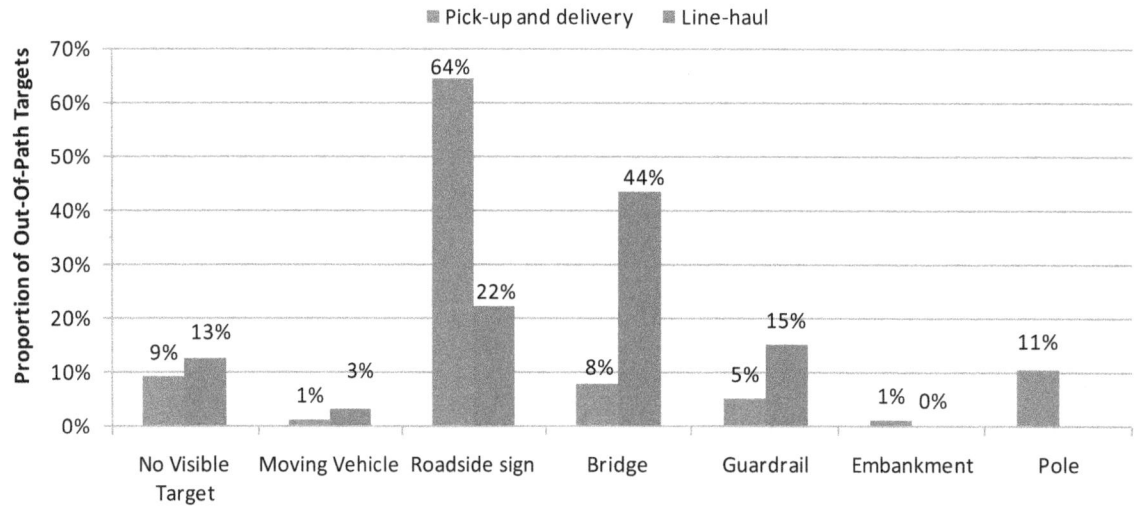

Figure 26. Distribution of out-of-path target types in FCW-S alerts

It should be noted that at night "no visible target" was coded in videos recorded when no lead vehicle was observed and it was too dark to identify the target type on either side of the road. Figure 27 provides the relative frequency of FCW alerts triggered by out-of-path targets as a result of host maneuvers, road position, and location. Out-of-path targets accounted for about 46 percent of all FCW alerts issued in construction zones, 43 percent at curve entry, and 43 percent when the host truck was negotiating a curve.

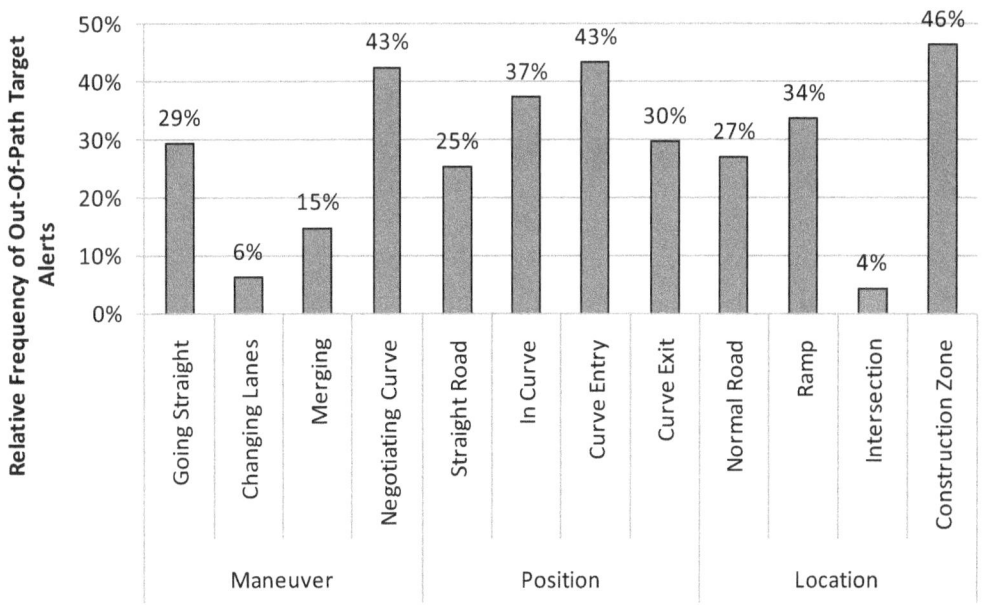

Figure 27. Relative frequency of FCW alerts triggered by out-of-path targets

4.1.2 Side-Looking Sensors

Performance of the side-looking sensors was characterized by examining the presence and relative location of adjacent targets for LCM and LDW-I alerts. This analysis was based on the review of 2,694 LCM alerts and 4,167 LDW-I alerts. Lane-change/merge warning systems are designed to detect objects in adjacent lanes that pose a threat during a lane-change maneuver or when the equipped vehicle is drifting out of its travel lane. Adjacent targets are any object occupying the lane adjacent to the vehicle. This includes the area directly adjacent to the equipped vehicle on both sides, or in the closing zone (in the adjacent lane, but behind the equipped vehicle). Vehicles in the adjacent lane ahead of the equipped vehicle do not pose a threat because they allow the driver enough room to safely make a lane change, while vehicles two or more lanes over do not pose a threat during a lane change. The percentage of LCM and LDW-I alerts issued when no targets were present in the adjacent lanes was used to measure system performance. The distribution of alerts issued for non-adjacent targets is provided by target type, host truck maneuver, position, and location.

About 42 percent of all LCM alerts and 44 percent of all LDW-I alerts were issued when a target was not present in the adjacent lanes. This is depicted in Figure 28. For line-haul drivers, the percentage was evenly divided between LCM and LDW-I alerts at approximately 50 percent. In contrast, about 25 percent of all LCM alerts and 37 percent of all LDW-I alerts received by pick-up and delivery drivers were attributed to no targets present in the adjacent lanes.

Figure 29 illustrates the distribution of LCM and LDW-I alerts due to non-adjacent targets by type for all drivers. In over 50 percent of all non-adjacent target alerts, reviewers did not observe any vehicle in adjacent lanes and were unable to identify the target. From the remaining cases, moving vehicles and roadside signs were the most frequent targets for pick-up and delivery drivers, while guardrails were the leading target for line-haul drivers. Of all LCM and LDW-I alerts attributed to observed targets, about 23 percent were issued in response to objects located two or more lanes over.

In 85 percent of all LCM non-adjacent target alerts, the host truck was changing lanes or merging On the other hand, the host truck was going straight in 82 percent and negotiating a curve in 12 percent of all LDW-I non-adjacent target alerts. About 5 percent of all LCM non-adjacent alerts occurred on exit ramps, while two percent occurred in construction zones. Similarly, 2 percent of all LDW-I non-adjacent target alerts occurred on ramps, and 4 percent in construction zones.

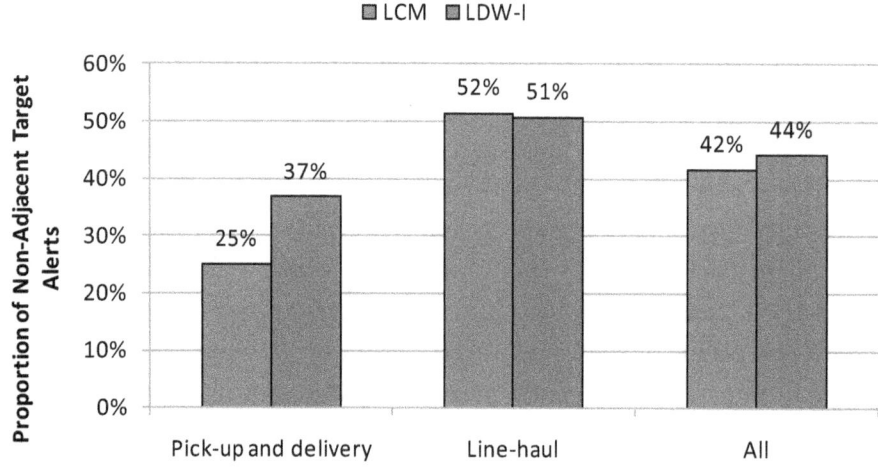

Figure 28. Proportion of LCM and LDW-I alerts triggered by targets not in adjacent lanes

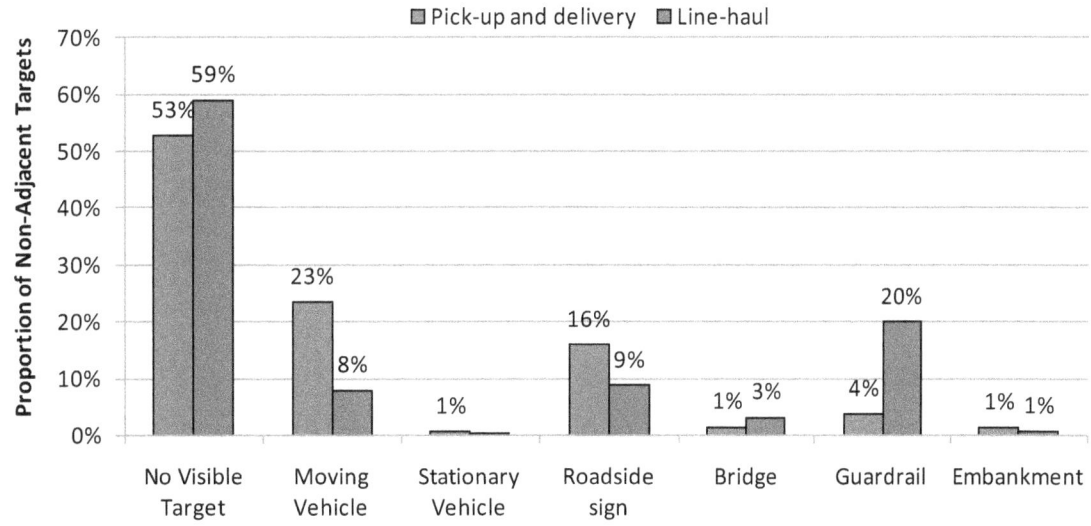

Figure 29. Distribution of non-adjacent target types for LCM and LDW-I alerts

Figure 30 provides the relative frequency of LCM and LDW-I alerts issued when no target was present in adjacent lanes, listed by host truck maneuver, road position, and location. Non-adjacent targets were associated with 44 percent of all LDW-I alerts issued when the host truck was going straight, as opposed to 26 percent of all LCM alerts. This rate was 48 percent of all LDW-I alerts issued when the host truck was merging, in contrast to 33 percent of all LCM alerts in this maneuver. Non-adjacent targets were found in 51 percent of all LCM alerts issued at curve exit, as opposed to 40 percent of all LDW-I alerts; this rate was 39 percent of all LCM alerts and 24 percent of all LDW-I alerts issued on ramps. The relative frequency of non-adjacent target alerts was observed in 27 percent of all LCM alerts and 51 percent of all LDW-I alerts at intersections.

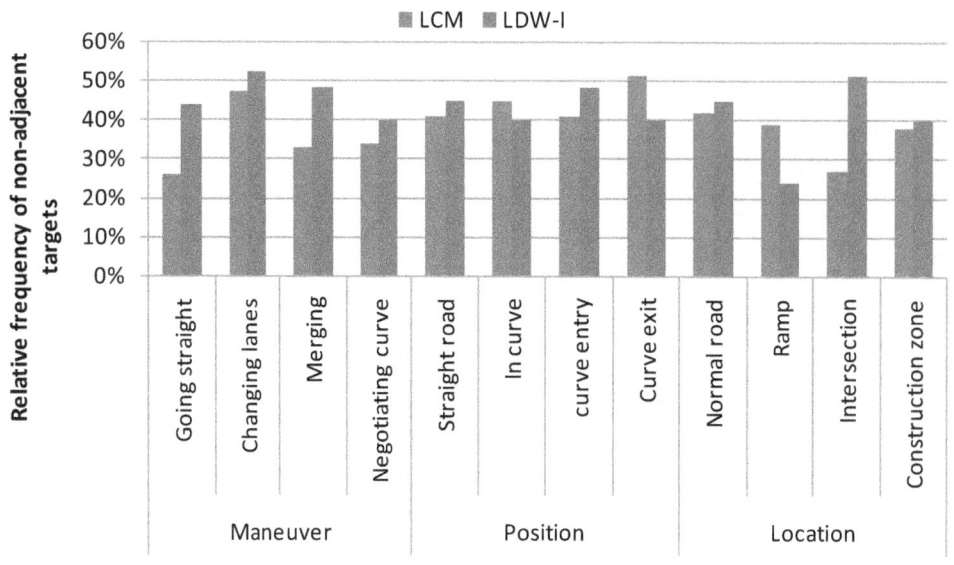

Figure 30. Relative frequency of LCM and LDW-I alerts triggered by non-adjacent targets

4.1.3 Lane Tracking

Performance of the vision-based lane-tracking system used for the LDW function was analyzed by its ability to detect unsignaled lane excursions. LDW-C warnings alert drivers when they are drifting out of their travel lane and no target is present. A sample of 3,671 LDW-C alert videos was analyzed to determine alert validity. This analysis distinguishes between unintentional and intentional lane excursions. Unintentional excursions result in alerts issued because the truck drifted over the lane marker, while intentional excursions account for alerts that were issued when drivers intentionally left their lane of travel (e.g., made a unsignaled lane change or maneuvered around an obstacle). All LDW-C alerts issued for intentional and unintentional excursions were considered to be valid. The main performance measure is the proportion of LDW-C alerts issued when the vehicle was not observed leaving its lane of travel. The distribution of these alerts is provided for road surface condition (based on video observation), lighting (using a system sensor), weather (wipers on), host truck maneuver and location.

Analysis of the lane tracking system revealed that for 10 percent of all LDW-C alerts issued, the host truck remained within its lane and did not cross a lane boundary. Figure 31 shows that 13 percent of all LDW-C alerts were issued when pick-up and delivery drivers remained in their lanes and did not cross a lane boundary, while only 7 percent of all LDW-C alerts were issued to line-haul drivers. Drivers received invalid LDW-C alerts (i.e., alerts that were not associated with any observable lane excursion) under these circumstances:

- 88 percent were issued while going straight, as opposed to 12 percent when merging or negotiating a curve.
- 85 percent on straight roads and 15 percent in curves, curve entries, or curve exits.

- 90 percent on roadways (arterials, limited-access highways or rural roads) as opposed to 10 percent on ramps, at intersections, or in construction zones.
- 93 percent on dry roads versus 7 percent on slippery road surfaces.
- 66 percent during the day and 34 percent at night.
- 95 percent in clear weather as opposed to 5 percent in adverse weather (wipers on).

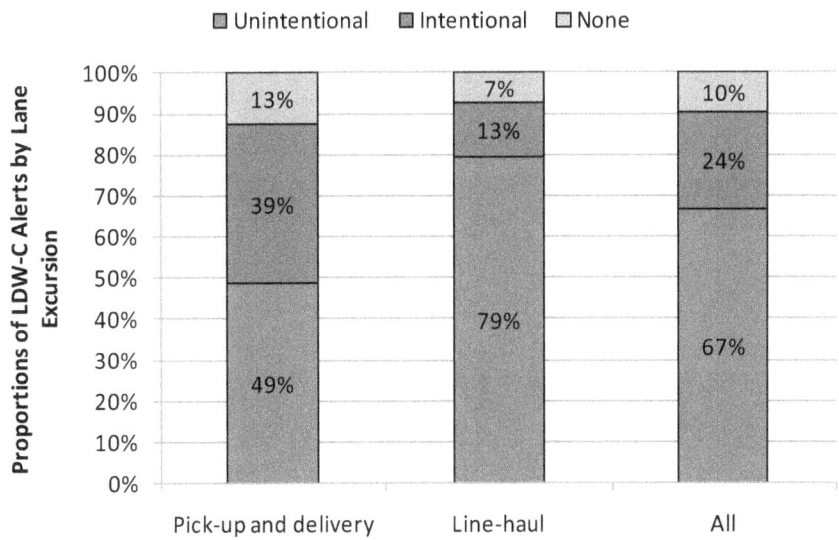

Figure 31. Distribution of LDW-C alerts by occurrence of lane excursion

Figure 32 illustrates the proportion of LDW-C alerts without any lane excursion under different driving conditions. For instance, for 22 percent of all LDW-C alerts issued at intersections, the vehicle did not leave its lane of travel, i.e., no lane excursion occurred. Furthermore, when the equipped vehicle was merging into another lane, 16 percent of all LDW-C alerts were invalid.

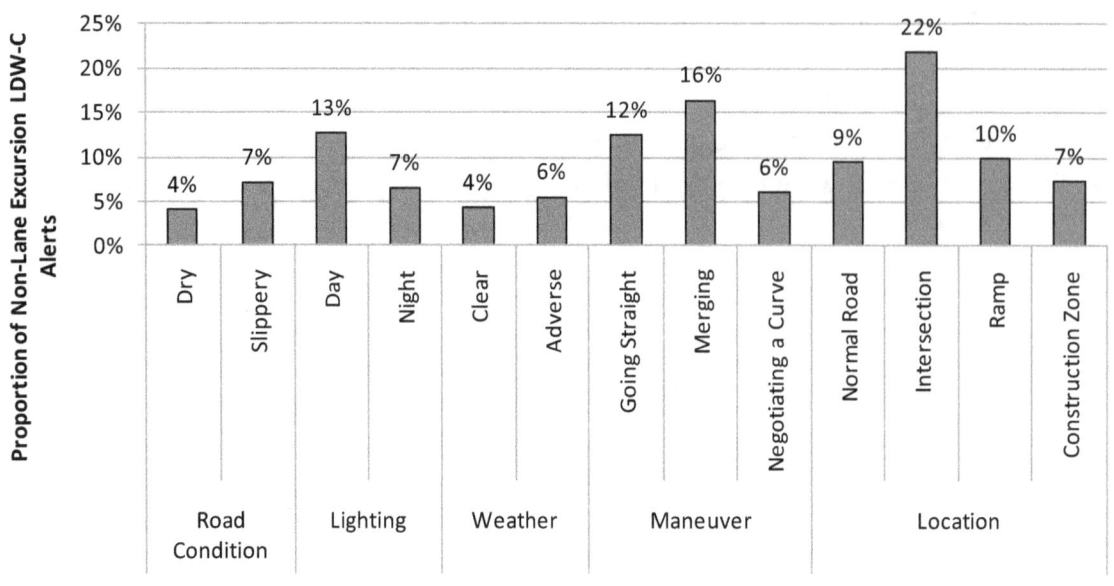

Figure 32. Proportion of LDW-C alerts without lane excursion under various conditions

4.2 Warning Logic

The performance of the warning logic was analyzed using a sample of alert episodes where an obstacle or road hazard was in the intended travel path of the host truck. Specifically, this analysis focuses on the in-path vehicle or obstacle category shown in Table 27. The "no hazard" and "out-of-path hazard" categories are addressed in the performance of sensors discussed above. A false alert is a warning caused by noise or interference when there is no object or threat present. Out-of-path nuisance alerts are caused by vehicles and objects that are not in the intended path of the host truck. In-path nuisance alerts refer to warnings for vehicles that are in the intended path of the host truck, but are at a distance or moving at a speed that drivers do not perceive as threatening. For instance, forward crash warnings are issued for lead vehicles turning at intersections. Some of these alerts could be issued based on the system design, but drivers usually perceive them to be unnecessary. In this section, the in-path vehicle or obstacle alerts are analyzed by hazard propensity and driver response.

Table 27. Analysis of system alerts

	No Hazard	In-Path Vehicle/Obstacle		Out-Of-Path Hazard
		Situation Requiring an Alert	Situation Not Requiring an Alert	
Alert issued	False alert	Appropriate alert	In-path nuisance alert	Out-of-path nuisance alert
No alert issued	Appropriate non-alert	Missed alert	Appropriate non-alert	Appropriate non-alert

Performance results of the warning logic are provided based on the analyses conducted on the hazard propensity and driver response to system alerts issued for valid crash threats.

4.2.1 Hazard Propensity

The efficacy of the warning logic to issue appropriate alerts is judged in two areas: mapping alerts to driving conflicts and near-crashes; and driver condition in the alert situation.

System alerts were mapped to the driving conflicts and near-crashes, as illustrated in Table 28. The parameters X1 and X4 represent a match between driving conflicts or near-crashes and system alerts. This match is determined by the overlap of the conflict duration over a time window ranging from 10 seconds before to 15 seconds after the onset of the alert. In contrast, the parameters X2, X3, X5, and X6 refer to mismatches between alerts and conflicts or near-crashes. Near-crashes form a more reasonable basis to assess where the integrated safety system might have issued an appropriate alert (X4), missed a hazardous situation (X5), or issued a nuisance alert (X6).

Table 28. Correlation between alerts and driving conflicts/near-crashes

Alert	Driving Conflicts		Near-Crashes	
	Yes	No	Yes	No
Yes	X1	X3	X4	X6
No	X2		X5	

Driver condition, a measure of involvement in secondary tasks or eyes-off-the-road, is also mapped to the alerts based on video observations recorded in the data logger. This analysis is based on a sample of alerts randomly selected from all driver trips. The rate of involvement in secondary tasks during alert episodes and rate of eyes-off-the-road alert episodes provide alternative measures for consistency, accuracy and reliability of the warning logic.

Figure 34 illustrates the mapping of valid alerts to driving conflicts when alerts were issued for three driver groups. Values for two measures that include the proportion of alerted conflicts over all driving conflicts and the proportion of alerted conflicts over all valid alerts are provided.

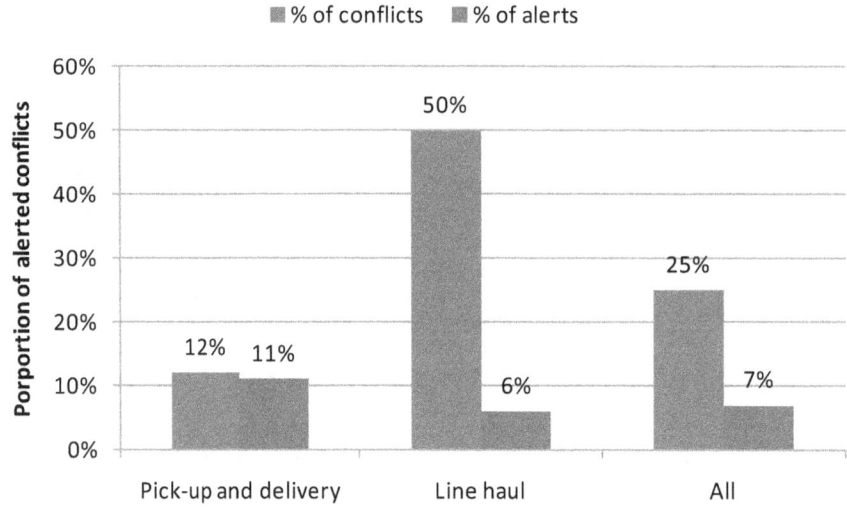

Figure 33. Mapping of valid alerts to driving conflicts

Figure 35 displays the results from a similar analysis conducted for near-crashes. It should be noted that the integrated system's warning logic does not necessarily match the definition of driving conflicts and near-crashes used in this report. Moreover, many conflicts and near-crashes did not result in alerts being issued. This is because the alerts may have been suppressed for one or more of the following reasons: turn signal use; braking; lane-change maneuvers; occurrence of an earlier system alert; or low confidence in system measures or parameters.

Line-haul drivers received alerts for about 50 percent of the driving conflicts and 61 percent of the near-crashes they experienced during the field test. In contrast, pick-up and delivery drivers received issued alerts in only 12 percent of the driving conflicts and 17 percent of the near-crashes they encountered. This discrepancy can be explained by the fact that pick-up and delivery drivers use their signals, brake, and make lane changes more often than line-haul drivers, due to more mileage driven on surface streets. A small percentage of all valid alerts were associated with driving conflicts and near-crashes.

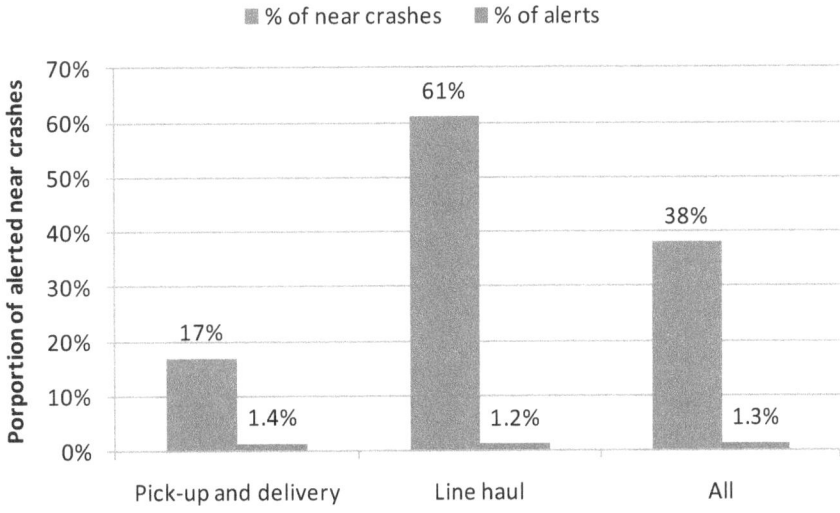

Figure 34. Mapping of valid alerts to near-crashes

Figure 35 presents the results of involvement in secondary tasks and the eyes-off-forward-scene condition prior to the onset of system alerts for valid threats. These results are based on video analysis of 9,398 alert episodes that consist of 23 percent FCW, 17 percent LCM, 25 percent LDW-I, and 35 percent LDW-C alerts. Not included in the list of secondary tasks is driver scanning of the blind spots on either side of the truck. However, this activity is included in the determination of eyes-off-forward-scene that involves any extended time prior to an alert when the driver was not looking at the forward scene. The highest rates of secondary tasks (66%) and eyes-off-forward-scene (19%) were observed in LDW-C alerts. FCW alerts had the lowest percentage of eyes-off-forward-scene at six percent.

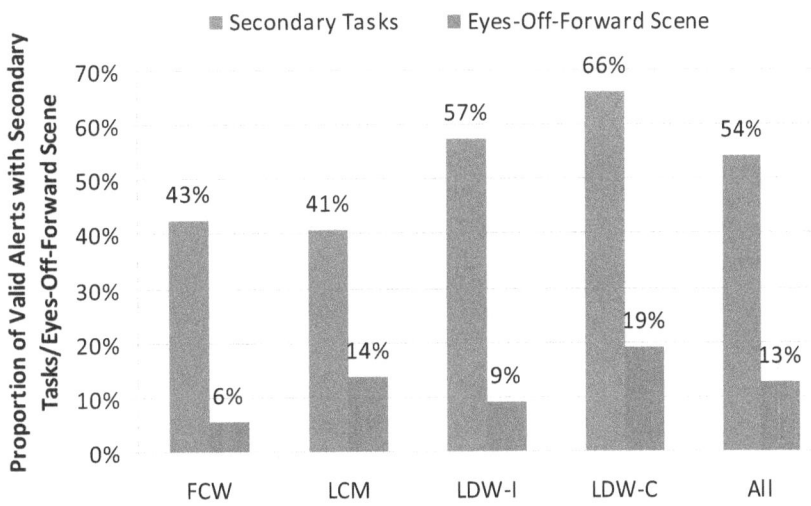

Figure 35. Proportion of valid alerts associated with secondary tasks and eyes-off-forward-scene

4.2.2 Driver Response

Results of driver response to system alerts were derived from subjective and objective data. Eight questions in the post-drive survey located in Appendix A addressed drivers' assessment of the appropriateness of system alerts. Four questions addressed the frequency of false alerts and the other four solicited drivers' reaction to false alerts. In these questions, the term "false" alert was based on the interpretation of the driver taking the survey; therefore, it refers to any alert that the driver did not feel was necessary.

Figure 36 illustrates drivers' responses to the unnecessary system alerts, including annoyance with nuisance alerts and the alert's effect on their trust and understanding of the system. (Note that for the latter two questions, disagreeing with the statement indicated a favorable opinion of the system.) The following are findings from the driver survey questions:
- Half of the 18 drivers felt that nuisance alerts did not affect them negatively;
- Seven drivers were annoyed by false alerts;
- Eight drivers agreed that the number of false alerts caused them to ignore the system warnings; and
- Five reported that false alerts affected their ability to understand the system properly.

One of the line-haul drivers commented that the false alerts were especially annoying when they occurred repeatedly for the same target, such as an overhead bridge or during a rainstorm.

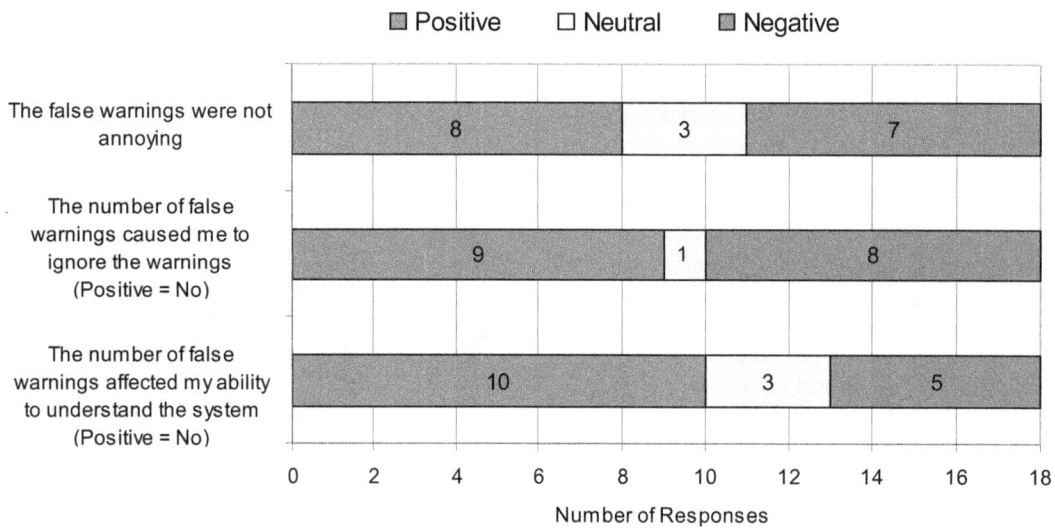

Figure 36. Responses to survey items about drivers' reaction to nuisance alerts

In response to the open-ended question "How did the false warnings affect your perception of the integrated system?" line-haul drivers were more annoyed by the false alerts than pick-up and delivery drivers. Responses included the following:
- "When you heard or saw a warning, you automatically checked what was happening around you, this was good to do." (pick-up and delivery driver)
- "[The false alarms were] not a big issue, bridges and guardrails set it off the most." (pick-up and delivery driver)
- "Sometimes I felt [the system] did not give me enough space in front of me, but I would rather have too much warning than not enough." (pick-up and delivery driver)
- "I felt the system just needed some more fine-tuning." (pick-up and delivery driver)
- "Some of the warnings were false more often than not, and it caused me to ignore some of them." (line-haul driver)
- "A lot were in the same places, i.e., bridge, so you got used to it." (line-haul driver)
- "I had less confidence in the whole system [because of the false warnings]." (line-haul driver)
- "[The false alerts] made it hard for me to trust the system." (line-haul driver)
- "The system still had bugs that needed to be worked out." (line-haul driver)

Figure 37 shows drivers' perception of each type of false alert. Drivers were asked the same question for the system overall, as well as for each type of alert, which allows for comparison of responses across alert types. Fifteen drivers agreed that the system issued alerts when they did not need them. These results show that drivers thought drift alerts were issued more appropriately than side hazard and front hazard alerts.

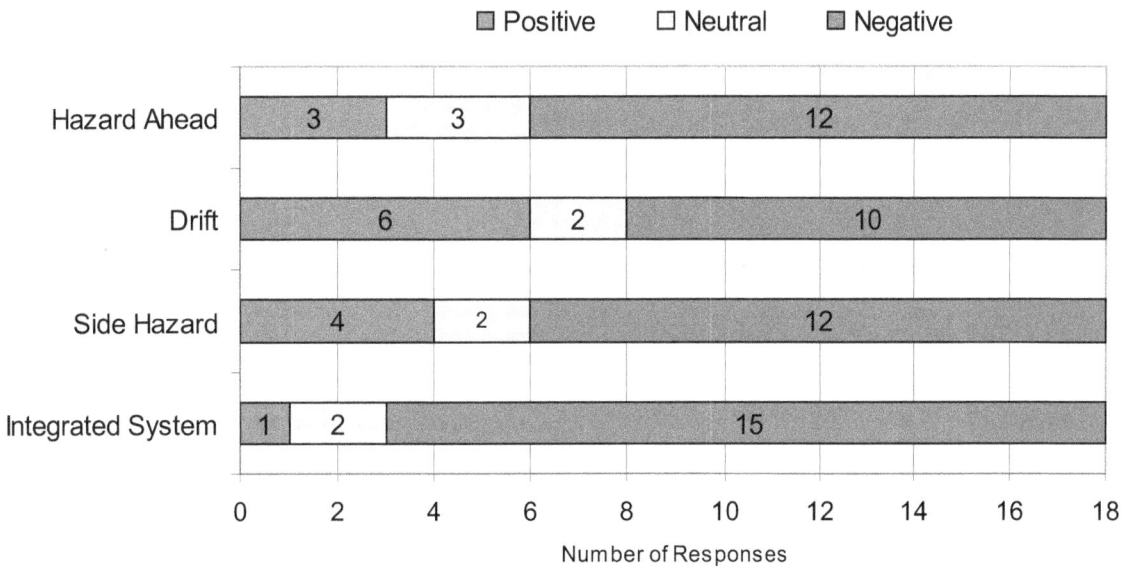

Figure 37. Drivers' opinions of false warnings by alert type

In addition to subjective assessment of the warning logic, data analysis was conducted to objectively infer whether or not system alerts impacted driver performance. In alert episodes where a hazard existed in the path of the host truck, driver response to the alerts was compared between the baseline and treatment periods. Driver response was expressed in terms of response type, brake reaction time and peak deceleration level to FCW alerts, and peak lateral acceleration to lateral alerts.

Figure 38 illustrates the action taken by drivers in response to alerts issued for valid threats. Drivers appeared to respond at a slightly higher rate during the treatment period than during the baseline period in response to FCW alerts (36 percent versus 32%) and LDW-C alerts (25 percent versus 20%). In the treatment period, a higher rate of braking (32%) was observed during FCW alerts and a higher steering rate (19%) was observed during LDW-C alerts. The intensity of response appears to be unchanged between baseline and treatment periods for LCM and LDW-I alerts.

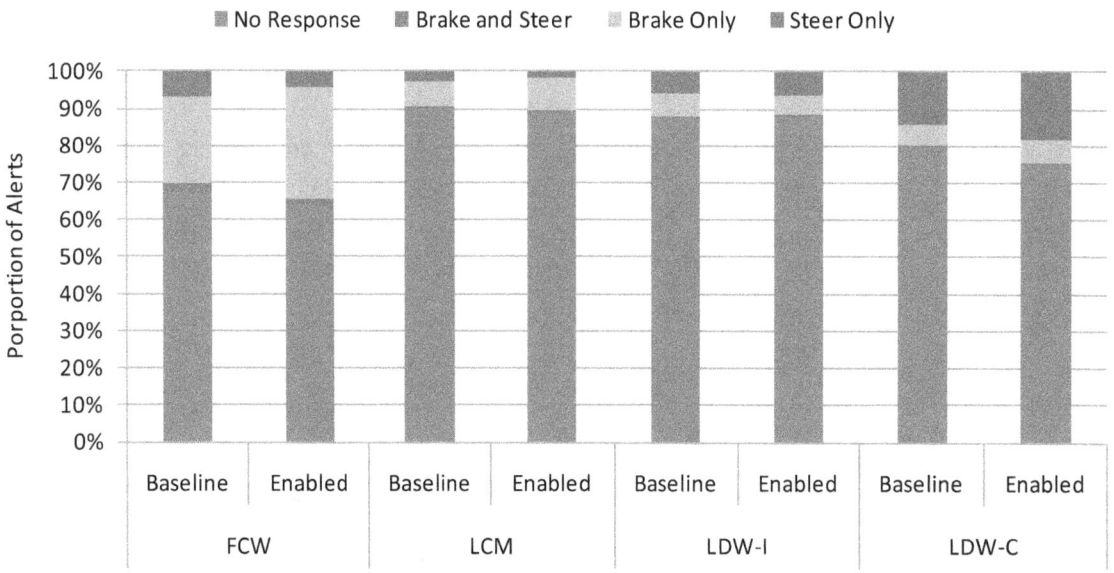

Figure 38. Breakdown of driver action in response to valid alerts

Drivers braked more quickly in response to FCW alerts, at an average of 1.50 seconds following alert onset during the treatment period, compared to 1.56 seconds during baseline driving. This minor difference is not statistically significant, as shown in Figure 39 (error bars represent the 95 % confidence interval). Brake reaction time to FCW alerts was measured from the time of alert onset until the time of brake pedal application. There was also no significant change in the intensity of braking action in response to FCW alerts between the baseline and treatment test conditions as shown in Figure 40. Braking response was measured by the peak deceleration level for each braking event. The average value was 0.07g (g = 9.81 m/s^2) in both test conditions, suggesting a mild braking response to FCW alerts.

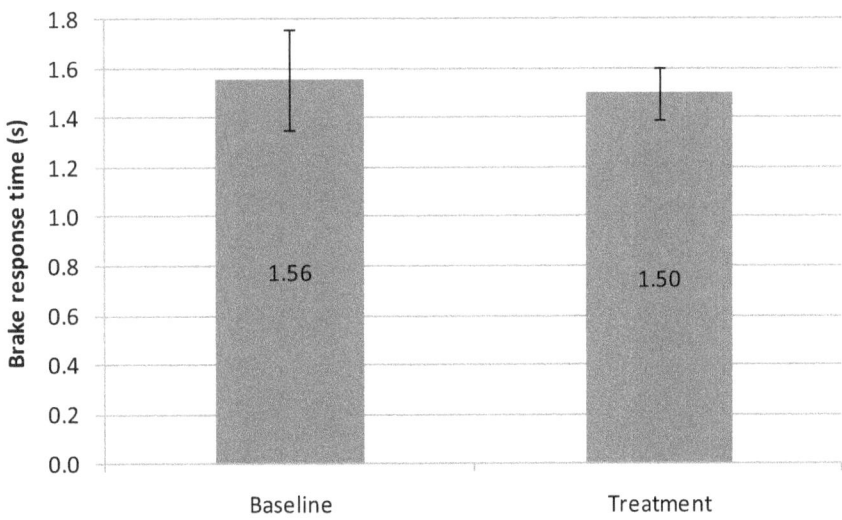

Figure 39. Average brake reaction time to FCW alerts between baseline and treatment

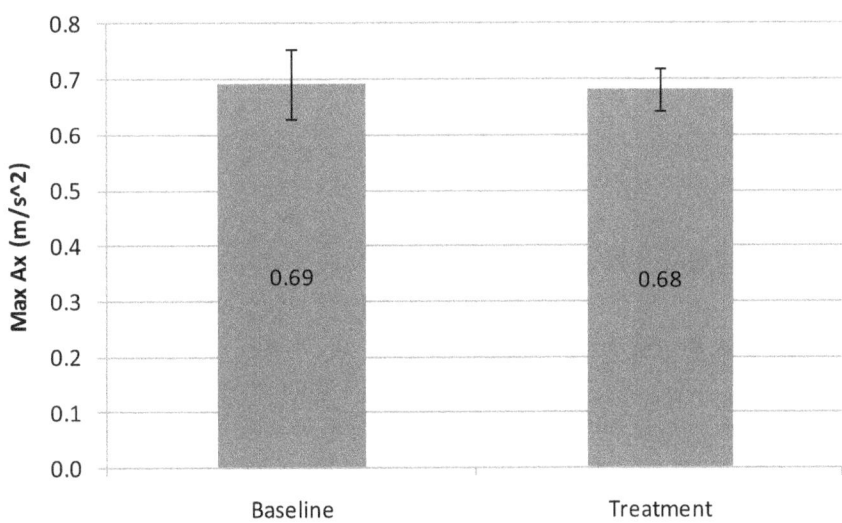

Figure 40. Average peak deceleration level to FCW alerts between baseline and treatment

Figure 41 shows that drivers reacted with higher peak lateral acceleration levels in treatment than during baseline driving in response to LCM, LDW-I, and LDW-C alerts. This increase in the intensity of steering response is statistically significant for LCM alerts and all side alerts combined. Drivers steered at an average peak lateral acceleration of 0.08g in response to LCM alerts in the treatment condition as opposed to 0.06g during baseline driving. In response to all side alerts combined, the average peak lateral acceleration level was 0.08g in the baseline condition and 0.07g during treatment. In valid threat situations, the drivers took a stronger corrective steering action in response to the LCM, LDW-I, and LDW-C alerts issued in the treatment period than during the baseline condition when auditory alerts were suppressed.

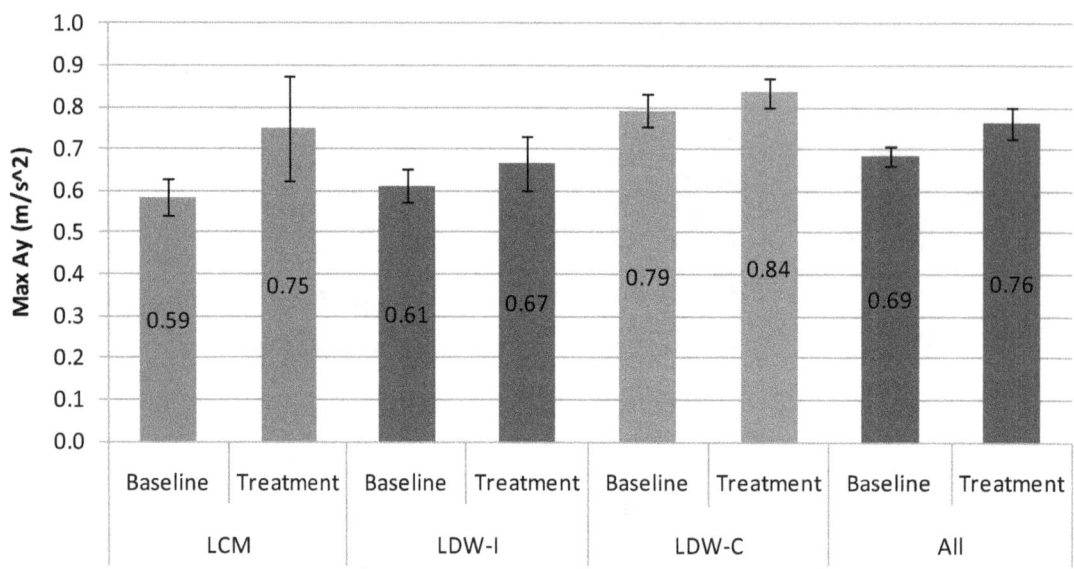

Figure 41. Average peak lateral acceleration to side alerts between baseline and treatment

4.2.3 Comparison of Alert Rates between Baseline and Treatment

Table 29 compares the alert rates from the four subsystems between the baseline and treatment conditions using paired t-test statistics. In this table, the average numbers of individual alert types per 100 miles traveled are presented along with the two-tail p statistic. This measure points to the statistical significance of the difference observed between the baseline and treatment conditions. The difference in LDW-C alerts between the two conditions is statistically significant at confidence levels over 94 percent for each of the three driver categories. The LDW-C alert rates decreased by 14 percent for pick-up and delivery drivers, 25 percent for line-haul drivers, and 20 percent for all drivers, from baseline to treatment. The cautionary alerts for lateral drift into unoccupied zone appear to impact the lane keeping behavior of test subjects who maintained their truck close to the lane center and thus reduced the frequency of LDW-C alerts. Moreover, the higher use of turn signals might have also contributed to the decline of LDW-C alert rates in the treatment condition.

Table 29. Average number of alerts per 100 miles driven in baseline versus treatment

Route Type	FCW			LCM			LDW-I			LDW-C		
	B	T	p	B	T	p	B	T	p	B	T	p
Pick-up and delivery	6.8	6.5	0.57	3.5	2.9	0.37	9.2	8.3	0.49	6.9	6.0	**0.02**
Line-haul	2.2	2.5	0.34	1.9	1.9	0.95	7.6	7.0	0.26	7.0	5.2	0.06
All	4.2	4.3	0.76	2.6	2.4	0.39	8.3	7.6	0.22	7.0	5.6	**0.01**

Table 30 compares the alert rates from three alert types (FCW-M, LDW-C left, and LDW-C right) between the baseline and treatment test conditions using paired t-test statistics. These alert types were issued in more valid situations than any other alert types. As shown in Figure 25, about 93 percent of all FCW-M alerts were issued for in-path targets. Similarly, Figure 31 indicates that about 90 percent of the LDW-C alerts were issued by lateral drift or an intentional maneuver to cross a lane boundary without the use of turn signals. Thus, the FCW-M and LDW-C alerts provide valuable insight into driver encounters with safety critical situations in baseline and treatment test conditions. The reduction of these alert rates from baseline to treatment implies potential safety benefits from use of the integrated system. Considering that the FCW-M and LDW-C alerts provide a surrogate indicator for exposure to driving conflicts in the field test, the scenario exposure ratio (ER) defined in Section 2.5 can be estimated for rear-end lead vehicle moving and lead vehicle decelerating, road-edge departure/no maneuver, and opposite direction/no maneuver driving conflicts.

Table 30. Average number of alerts per 100 miles driven in baseline versus treatment for FCW-M, LDW-C left, and LDW-C right alert types

Route Type	FCW-M					LDW-C Left					LDW-C Right				
	B	T	p	T4	p	B	T	p	T4	p	B	T	p	T4	p
Pick-up and delivery	6.1	5.5	0.08	5.6	0.39	3.8	3.1	**0.04**	3.0	0.06	3.1	2.9	0.20	2.7	0.10
Line-haul	0.96	0.73	**0.03**	0.67	**0.05**	3.1	2.3	0.12	1.9	0.09	3.9	2.9	0.09	2.6	0.12
All	3.3	2.9	**0.01**	2.8	0.15	3.4	2.7	**0.01**	2.4	**0.02**	3.5	2.9	**0.05**	2.6	**0.05**

As seen in Table 30, the FCW-M alert rate decreased from baseline to treatment for pick-up and delivery drivers (10 % reduction at 92 % confidence level), line-haul drivers (24% drop at 97 % confidence level), and all drivers (12 % decline at 99% confidence level). Based on extended system use, the test subjects appeared to improve their driving behavior as indicted by a lower FCW-M alert rate during the fourth period of the treatment condition.

Figure 42 illustrates the average reduction of the FCW-M alert rates between the baseline and treatment conditions, along with the 95 percent confidence bounds. For all drivers, the average reduction of alerts ranged from 6 to 27 percent between the baseline and treatment and from 3 to 35 percent between the baseline and the fourth period of the treatment condition at the 95 percent confidence bounds. This statistically-significant drop in FCW-M alert rates could be attributed to information the drivers received from the time headway display during the treatment test condition.

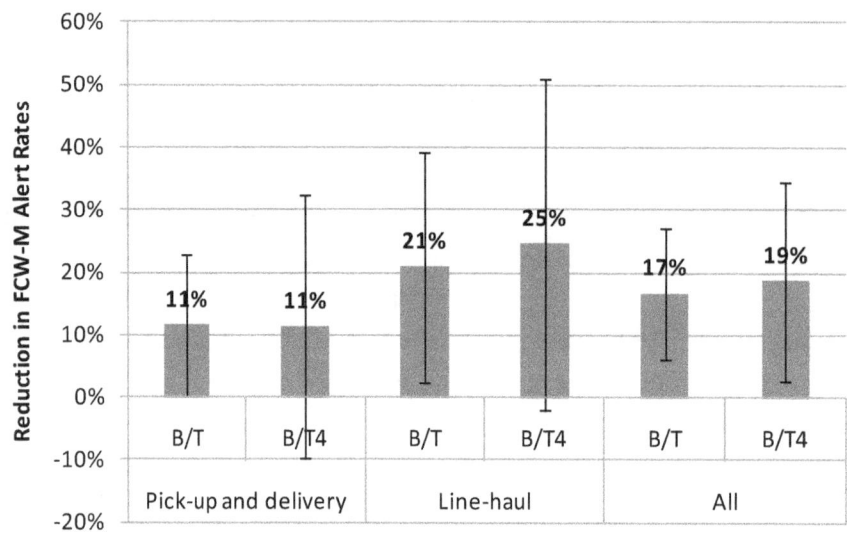

Figure 42. Change in FCW-M alert rates between test conditions

The LDW-C alert rates for lateral drifts in both directions also decreased during the treatment period, as illustrated in Table 30. The LDW-C alert rates for left lateral drifts were 18 percent less for pick-up and delivery drivers (96 percent confidence level), 25 percent less for line-haul drivers (88 percent confidence level), and 21 percent less for all drivers (99 percent confidence level). For lateral drifts to the right, LDW-C alert rates decreased by 8 percent for pick-up and delivery drivers (80 percent confidence level), 26 percent for line-haul drivers (91 percent confidence level), and 19 percent for all drivers (95 percent confidence level). Drivers maintained a better lane-keeping behavior throughout the entire treatment period, as evidenced in the reduction observed in the fourth period.

The average reduction of the LDW-C alert rates for lateral drifts in both directions (at a 95% confidence level) is shown in Figures 43 and 44. The average reduction of LDW-C alerts for lateral drifts to the left ranged from 6 to 29 percent and from 8 to 39 percent between the baseline and the fourth treatment period. On the other hand, the average reduction of the LDW-C alerts for lateral drifts to the right was slightly higher than the LDW-C alerts to the left.

Using observed reductions in alert rates in the treatment condition as values for the scenario exposure ratio and assuming that the values of the crash prevention ratio (PR) were equal to one; full deployment of the integrated safety system could potentially prevent about 8,000 police-reported crashes involving at least one heavy truck annually. The 95 percent confidence bounds of this estimate range between 3,000 and 13,000 police-reported crashes annually.

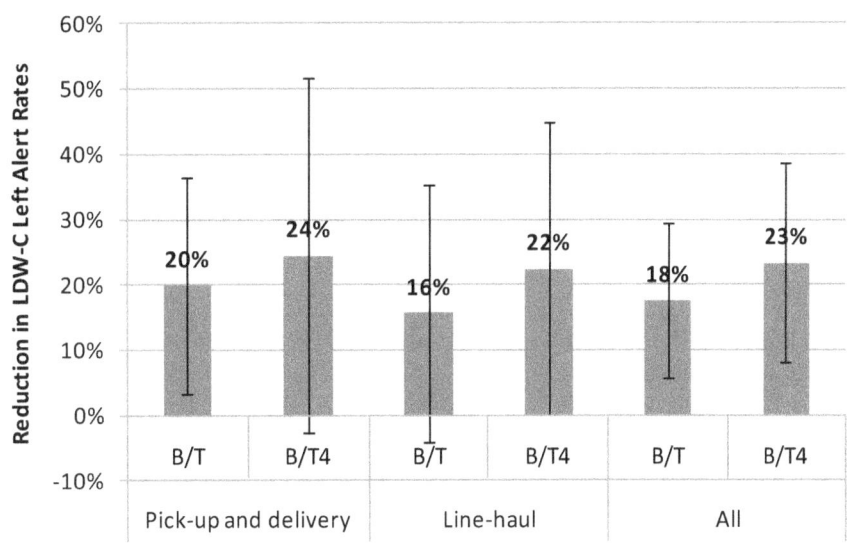

Figure 43. Change in LDW-C left alert rates between test conditions

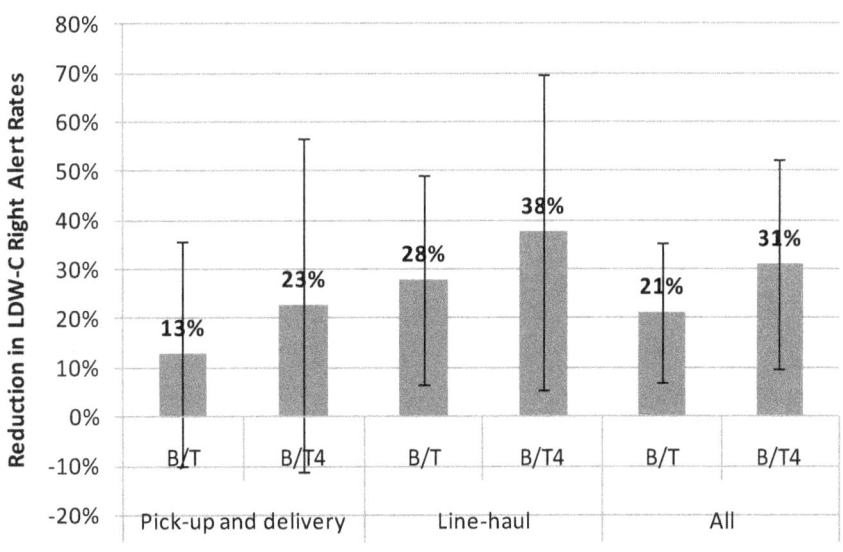

Figure 44. Change in LDW-C right alert rates between test conditions

4.3 Driver-Vehicle Interface

Analysis of the driver-vehicle interface focused on the system display, auditory warnings, and system controls. The driver vehcle interface includes the center display and blind spot monitor displays mounted on the the truck's A-pillars, as shown in Figure 1. Readability of visual information and the auditory alerts signals were evaluated through survey responses. Post-drive surveys (located in Appendix A) and subject debriefings were used to collect driver feedback.

Figure 45 shows the results of four survey items related to the usefulness of the displays. Slightly more than half of the drivers rated the system displays as being useful. Drivers commented that the display was easy to read, and that it was in a convenient location. Eleven drivers thought that the LDW availability icons on the center display were useful; two drivers said that they did not notice the icons. Eleven drivers found the blind spot monitor displays to be useful, while 5 said they did not find them to be useful. One of these drivers, a line-haul driver, commented that a visual warning was issued for guardrails that were more than 10 feet away. Finally, half of the drivers thought that the blind spot monitors were mounted in the proper locations, while the other half were neutral or thought that these displays should have been placed in another location. One of the drivers who thought the blind spot monitors were not placed in a convenient location requested that they be moved to the instrument panel, while another driver mentioned that the passenger side display was too far away and should be moved closer to the driver. Overall, the system displays were rated favorably and received only a minimal amount of negative responses from the drivers.

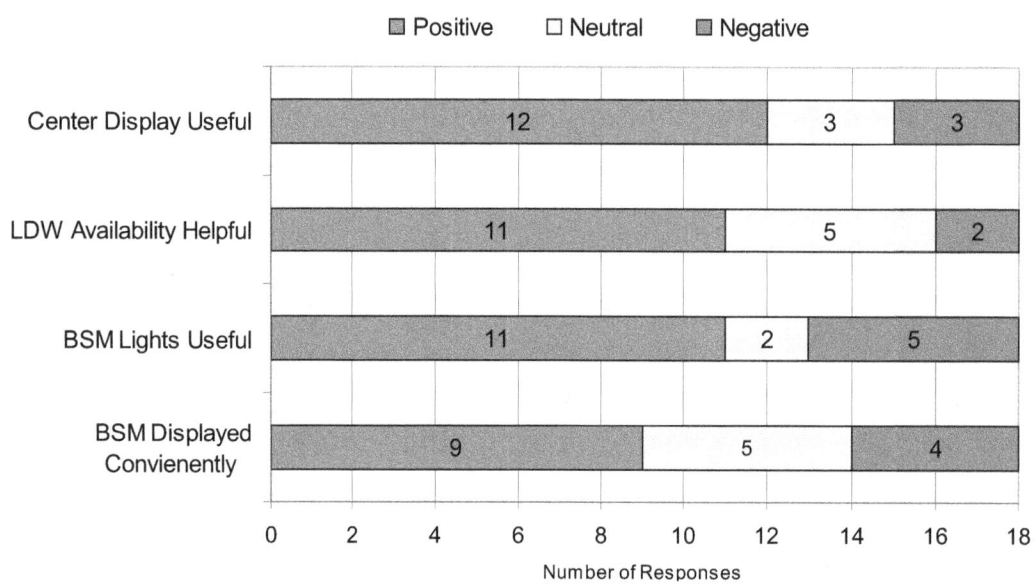

Figure 45. Drivers' opinions of the usefulness of the integrated system displays

Driver feedback indicated that most thought the auditory alerts were quite useful. They reported that they could easily distinguish between the two different warnings sounds, and also that the warnings were attention-getting. These results are illustrated in Figure 46.

The two system controls include a button to adjust the volume of the auditory alerts and a mute button to suppress the alerts for a period of up to six minutes. Use of the mute button allowed drivers to temporarily disable auditory alerts in areas such as construction zones. Figure 47 illustrates the number of times each driver used the mute button during the field test. Of the 18 drivers, only 2 drivers used the mute button on a regular basis. Both of these drivers reported that they found the mute button to be very useful. Driver 30 mentioned that he used the mute button in construction zones. The other 16 drivers did not take advantage of the mute function, most likely due to personal preference for ignoring alerts rather than suppressing them.

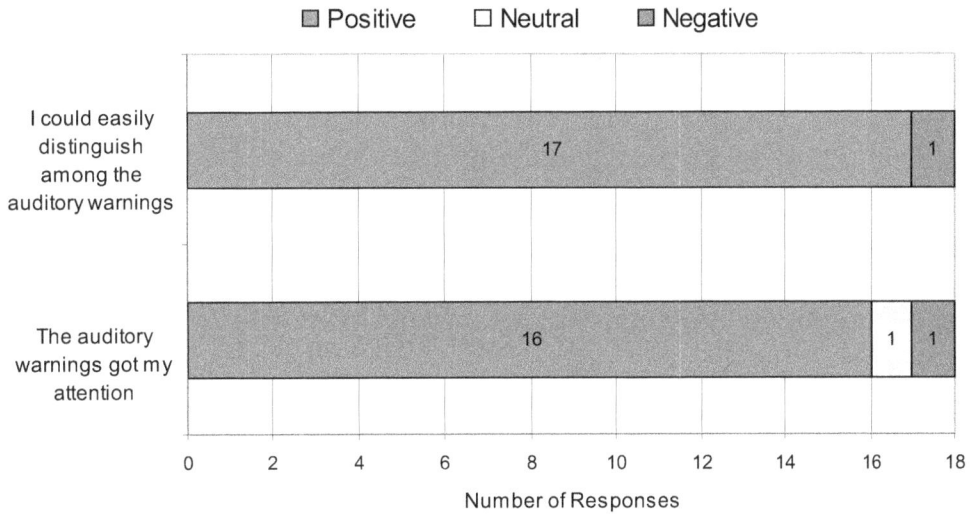

Figure 46. Drivers' opinions of the integrated system auditory warnings

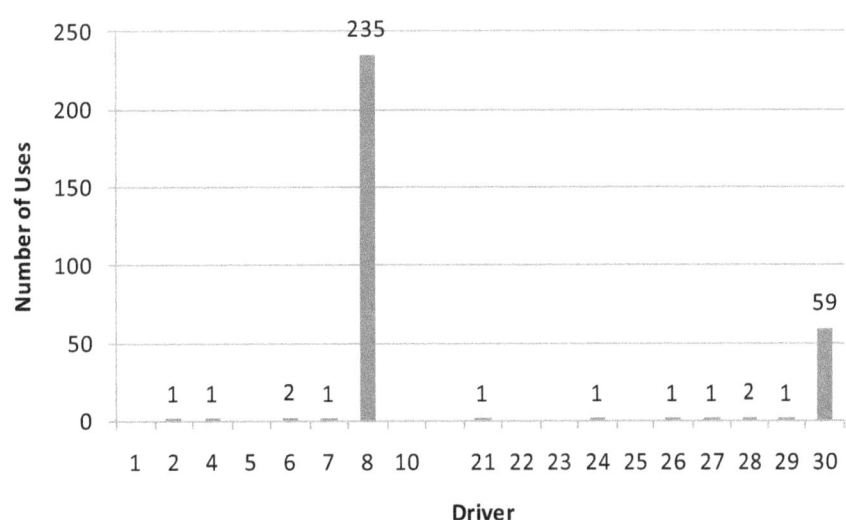

Figure 47. Number of mute button uses during the field test

Volume control use by each driver is shown in Figure 48. While a handful of drivers adjusted the volume a number of times, most drivers adjusted the volume fewer than 20 times during the field test. Drivers most likely found the volume setting they preferred when the system was initially enabled and did not find the need to change the volume setting. Only 4 drivers adjusted the volume more than 35 times.

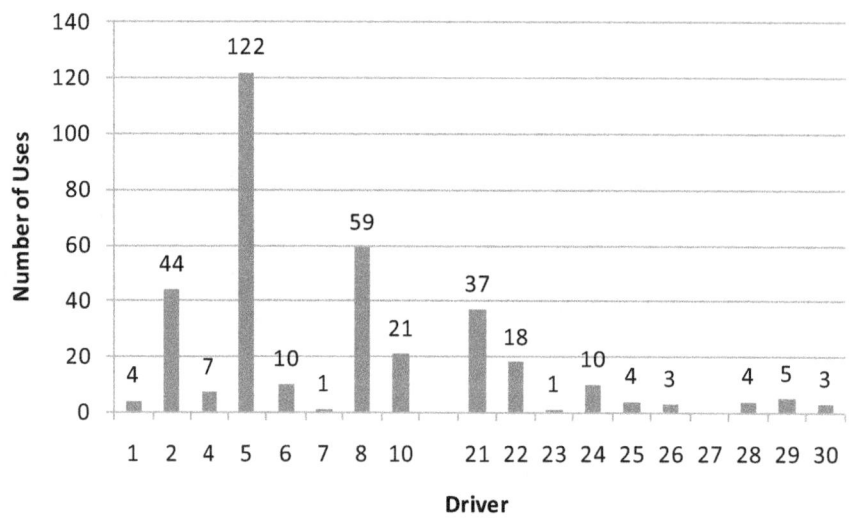

Figure 48. Number of volume adjustments during the field test

The subjective assessment of the usefulness of system controls is shown in Figure 49. Half of the drivers responded positively to the usefulness of the volume control, while the majority of the drivers did not find the mute button to be useful. The 2 drivers who regularly used the mute button both rated its usefulness a numerical rating of 7, the highest rating.

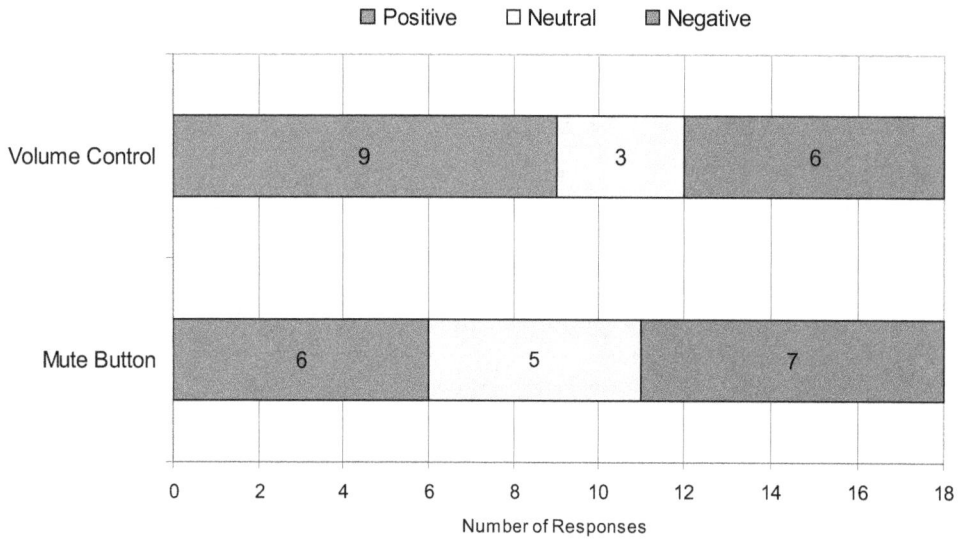

Figure 49. Drivers' opinions of the usefulness of system controls

4.4 System Robustness

System robustness was assessed by evaluating the availability of the integrated system's LDW function to issue a lane departure alert. System availability is a measure of its ability to recognize and track lane markers. It is important to note that the LDW function was disabled when the driver activated the turn signal or depresses the brake pedal. The LDW function was considered "available" when it was able to recognize and track both lane markers. This enabled the function to issue crash alerts for lateral drifting. Figure 50 illustrates the LDW availability under all driving conditions for left right, and both sides of the travel lane for three different speed ranges. The lowest speed range (vehicle speeds between 25 and 35 mph) represents driving on local and rural roads. The middle speed range (between 35 and 55 mph) corresponds to driving on arterial roads. The upper speed range (over 55 mph) represents limited-access highway driving. The LDW function was available when tracking markers on both sides of the lane for 85 percent of the mileage driven at speeds above 55 mph. This rate is higher than the other two speed ranges because limited-access highway lanes are generally better marked and maintained. The LDW availability function drops to 34 percent of the mileage traveled at lower speeds due to lower quality lane marking conditions on local and rural roads.

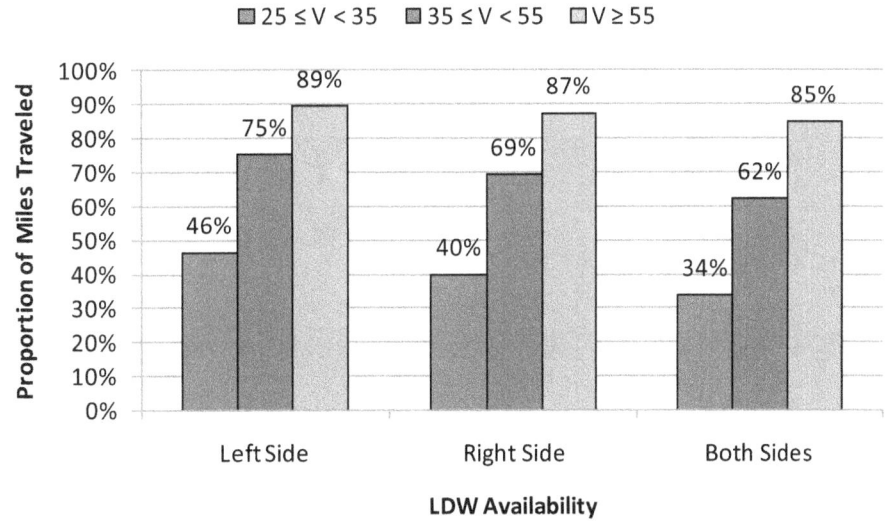

Figure 50. Availability of LDW function by travel speed

Table 31 summarizes availability of the LDW function by travel speed, ambient lighting and weather conditions. When tracking markers on both sides of the lane, the availability dropped by 22 percent at the lower speed range, 13 percent at the middle speed range, and 6 percent at the upper speed range from daytime to nighttime conditions. From clear to adverse weather conditions, this performance decreased by 20 percent at the lower speed range and 27 percent at the middle and upper speed ranges.

Table 31. Availability of LDW function by travel speed and driving conditions

Travel Speed (mph)	Left Side				Right Side				Both Sides			
	Lighting		Weather		Lighting		Weather		Lighting		Weather	
	Day	Night	Clear	Adverse	Day	Night	Clear	Adverse	Day	Night	Clear	Adverse
$25 \leq V < 35$	49%	38%	47%	38%	41%	36%	41%	30%	36%	28%	34%	27%
$35 \leq V < 55$	82%	64%	77%	57%	71%	66%	72%	52%	66%	57%	64%	47%
$V \geq 55$	96%	88%	92%	67%	92%	87%	90%	66%	90%	84%	87%	63%

5. Conclusions

This section presents the key findings and discussion of the independent evaluation of the heavy-truck field operational test including estimated crash reductions, changes in driver behavior, and driver's perceptions of the system and sensor accuracy.

Safety Benefits:
- Considering that forward-crash warning for moving targets and cautionary lateral-drift warning (alert types with a high rate of accuracy) are indicators of exposure to relevant driving conflicts, the full deployment of integrated safety systems could prevent between 3,000 and 13,000 police-reported truck crashes annually. The breakdown of this estimate by alert type and pre-crash scenario is shown in Table 32.

> **HIGHLIGHTS**
> - Potential safety benefits are projected for road-departure, opposite-direction, and rear-end crashes.
> - Unintended consequences included a slight increase in secondary tasks, but did not appear to negatively impact driving safety during the field test.
> - The majority of test subjects would prefer to drive a truck with the integrated safety system over a conventional truck.
> - The forward-crash warning function for moving vehicles and cautionary lateral-drift warning function were effective and reliable.
> - System changes are needed to improve reliability of side object classification of the LCM function and misclassification of stopped objects for the forward-crash warning function.

Table 32. Estimated safety benefits of the integrated system based on alert rate reduction

Function	Pre Crash Scenario	Annual Target Crashes	Maximum Estimated Crash Reduction	Maximum Estimated Effectiveness
FCW-M	Rear end/Lead vehicle decelerating Rear end/Lead vehicle moving	18,000	5,000	27%
FCW-S	Rear end/Lead vehicle stopped	19,000		
LCM	Changing lanes/same direction Turning/same direction	53,000	Insufficient data to estimate	
LDW-I	Drifting/same lane	7,000		
LDW-C Left	Opposite direction/No maneuver Road edge departure/No maneuver	11,000	3,000	29%
LDW-C Right	Road edge departure/No maneuver	15,000	5,000	36%
Integrated System	All	123,000	13,000	11%

- Line-haul drivers increased their use of turn signals when changing lanes when the system was enabled.

- Drivers experienced fewer road-departure near-crashes,[2] especially on the left side of the road, with the integrated system enabled. An 8-percent decline in the frequency of rear-end near-crashes was also observed. Twelve drivers experienced fewer near-crashes per 1,000 miles traveled. These results indicate that the system helped most drivers avoid a near-crash. Eight drivers reported that the integrated system prevented them from getting into a crash or near-crash.
- Drivers initiated hard steering in response to valid side-hazard and lateral-drift alerts, indicating that the alerts were effective in getting the driver's attention.
- Drivers received fewer forward-crash alerts for moving targets and cautionary lateral-drift alerts, indicating less exposure to rear-end, road-departure, and opposite-direction driving conflicts during the treatment period.
- Eight line-haul drivers were involved in slightly more secondary tasks (eating, drinking, and talking on a cellular telephone) as measured by the percentage of viewed alert videos with secondary tasks. This result did not appear to negatively influence driving safety, since lower rates of lane excursions were observed.

Driver Acceptance:
- The majority of drivers would prefer to use a truck with the integrated system over a conventional, unequipped vehicle and would recommend that their employer purchase the integrated system for their vehicle fleet.
- Drivers felt that the system increased their driving safety and made them more aware of their surroundings.
- Drivers found the system to be easy to learn and use, and found the auditory alerts to be easy to understand.
- Due to higher mileage and longer system exposure, line-haul drivers were more likely than pick-up and delivery drivers to experience false warnings, and were more likely to report annoyance with the system.

System Capability:
- There was a significant difference in system reliability for classifying in-path and out-of-path targets. For example, 97 percent of FCW alerts associated with stationary objects (primarily roadside signs and overhead bridges) were issued when no valid threat appeared to be present. On the other hand, only 7 percent of FCW alerts for moving objects were issued when no valid threat was present, indicating a high degree of accuracy of the forward radar classifying moving targets.
- Over 50 percent of side-hazard alerts were issued with no target present in the detection zone. The majority of these alerts could be attributed to the radar reflections off the trailer body (Sayer et al., 2010).

---

[2] Driver responses above a certain intensity level (for example, hard braking or steering) as defined in Section 2.4

- In 90 percent of the LDW-C alerts, the equipped truck was observed crossing the lane boundary without the use of turn signals, indicating high system accuracy. Lane drift alerts had the lowest reported instances of unnecessary or false alerts among the three alert types.
- The LDW function was available to issue alerts in 81 percent of the total mileage driven at speeds greater than or equal to 25 mph, meeting lane-tracking system performance requirements.
- Almost all (17 out of 18) drivers indicated that they could easily distinguish among the three different types of auditory warnings, and 16 drivers said that the auditory warnings were attention-getting. These results indicate that the auditory warnings were salient.

Overall, driving with the integrated system improved driver's performance by increasing their awareness of traffic around them and the position of their vehicle in their travel lane. The system encouraged better lane-keeping behavior by alerting them when they were drifting out of their lane, and to avoid potential rear-end crashes by letting them know when they were closing in or approaching a lead vehicle too closely. These features increased driver's awareness of their driving habits and helped improve their vigilance. Lateral-drift cautionary alerts reminded drivers to use their turn signal more regularly, a habit conducive to safe driving.

Drivers who participated in the study had favorable opinions of the system and thought that it would improve their driving safety. Since safety was a company-wide priority, almost all drivers said they would prefer driving a truck with the integrated safety system than a standard, unequipped truck. Most drivers were aware of the system's shortcomings and reported receiving false warnings and being annoyed by them, at least on occasion. Despite the system's shortcomings, drivers maintained their favorable view and desire to use the integrated system as a means of increasing their driving safety.

One factor that potentially prevented the system from showing a more notable safety benefit was the quality of the drivers participating in the study. Con-way Freight puts a priority on safety and the drivers involved in this study were very safety conscious, took pride in their driving and were generally vigilant about their driving habits. A more pronounced safety increase may have been shown had the drivers participating in the study had more room for improvement.

Poor reliability of side object classifiction and consistent misclassification of stopped objects in the vehicle's forward path were two shortcomings of the prototype system in need of improvement.

References

Ference, J.J., Szabo, S., & Najm, W.G. (2006). Performance Evaluation of Integrated Vehicle-Based Safety Systems. Proceedings of the Performance Metrics for Intelligent Systems (PerMIS) Workshop. Gaithersburg, MD: National Institute of Standards and Technology.

Frantz, Gary. (2010). "Out of the lab and Into the Cab: Con-Way Freight Goes High-Tech for Safety." Con-Way Web site, retrieved August 24, 2010: http://www.con-way.com/en/about_con_way/newsroom/press_releases/Jun_2010/2010_jun_29/

Lam, A. H., Bailin, A. Najm, W. G., & Nodine, E. E. (2009). Identification of Driving Conflicts in Heavy Truck Field Operational Test Data. Project Memorandum. (HS22A1). Cambridge, MA: Volpe National Transportation Systems Center.

Najm, W. G., & Smith, J. D. (2007). Development of Crash Imminent Test Scenarios for Integrated Vehicle-Based Safety Systems (IVBSS). (DOT HS 810 757). Washington, DC: National Highway Traffic Safety Administration.

Najm, W. G., Stearns, M. D., Howarth, H., Koopmann, J., & Hitz, J. (2006). Evaluation of an Automotive Rear-End Collision Avoidance System. (DOT HS 810 569). Washington, D.C: National Highway Traffic Safety Administration.

Sayer, J., Bogard, S., Funkhouser, D., LeBlanc, D., Bao, S. Blankespoor, A., Buonarosa, M.L., & Winkler, C. (2010). Integrated Vehicle-Based Safety Systems Heavy-Truck Field Operational Test Key Findings Report. Washington, DC: National Highway Traffic Safety Administration. In press.

Sayer, J., LeBlanc, D., Bogard, S., Nodine, E., & Najm, W. G. (2009). Integrated Vehicle-Based Safety Systems Third Annual Report. (DOT HS 811 221). Washington, DC: National Highway Traffic Safety Administration.

Appendix A: Post-Drive Survey

Subject # _____

Date _____

IVBSS Heavy-Truck Field Operational Test – Questionnaire and Evaluation

Please answer the following questions about the Integrated Vehicle Based Safety System (IVBSS). If you like, you may include comments alongside the questions to clarify your responses.

Example:

A.) Strawberry ice cream is better than chocolate.

```
    1        2        3        4        5        6        7
Strongly                                              Strongly
Disagree                                               Agree
```

If you prefer chocolate ice cream over strawberry, you would circle the "1," "2" or "3" according to how strongly you like chocolate ice cream, and therefore disagree with the statement.

However, if you prefer strawberry ice cream, you would circle "5," "6" or "7" according to how strongly you like strawberry ice cream, and therefore agree with the statement.

If a question does not apply:

Write "NA," for "not applicable," next to any question which does not apply to your driving experience with the system. For example, you might not experience every type of warning the questionnairesurvey addresses.

The integrated system consists of three functions. Please refer to the descriptions below as you answer the questionnaire.

Forward Collision Warning (FCW) – The forward collision warning function provided an auditory warning whenever you were approaching the rear of the vehicle in front of you and there was potential for a collision. When you received this type of warning, the display read "Collision Alert." Additionally, this system provided you with headway information in the display as you approached the rear of a vehicle (e.g., object detected, 3 seconds)

Lane Departure Warning (LDW) – The lane departure warning function provided an auditory warning whenever your turn signal was not on AND you were changing lanes or drifting from your lane. When you received this type of warning, the display read "Lane Drift" and a truck in the display appeared to be crossing a lane line.

Lane-Change/Merge Warning (LCM) – The lane-change/merge warning function provided an auditory warning whenever there was a vehicle in the truck's blind spot, your turn signal was on, and the system detected sideways motion indicating your intention to make a lane change. A red LED illuminated in the side display on whichever side you were making the lane change. Additionally, if your turn signal was off, and there was no indication that you were intending to make a lane change, but there was a vehicle in the truck's blind spot, a yellow LED was illuminated in the side display.

General Impression of the Integrated System

1. What did you like most about the integrated system?

   _____

2. What did you like least about the integrated system?

   _____

3. Is there anything about the integrated system that you would change?

   _____

4. How helpful were the integrated system's warnings?

   | 1 | 2 | 3 | 4 | 5 | 6 | 7 |

   Not all   Very
   Helpful   Helpful

5. In which situations were the warnings from the integrated system helpful?

   _____

6. **Overall, I think that the integrated system is going to increase my driving safety.**

   1   2   3   4   5   6   7

   Strongly Disagree   Strongly Agree

7. **Driving with the integrated system made me more aware of traffic around me and the position of my truck in my lane.**

   1   2   3   4   5   6   7

   Strongly Disagree   Strongly Agree

8. **How long after it became enabled did it take you to become familiar with the operation of the integrated system (a day, a week, etc.)**

   _____

9. **The integrated system made doing my job easier.**

   1   2   3   4   5   6   7

   Strongly Disagree   Strongly Agree

10. **Did the integrated system prevent you from getting into a crash or a near-crash?**

    Yes_____ No_____

    If Yes, please explain _____

11. **I was not distracted by the warnings.**

    1   2   3   4   5   6   7

    Strongly Disagree   Strongly Agree

12. Overall, how satisfied were you with the integrated system?

| 1 | 2 | 3 | 4 | 5 | 6 | 7 |

Very
Dissatisfied
　　　　　　　　　　　　　　　　　　　　　　　　　　　　Very
　　　　　　　　　　　　　　　　　　　　　　　　　　　　Satisfied

13. Did you rely on the integrated system? Yes____ No____

　　a. If yes, please explain?

_____

　　As a result of driving with the integrated system did you notice any changes in your driving behavior? Yes____ No____

　　b. If yes, please explain.

_____

I always knew what to do when the integrated system provided a warning.

| 1 | 2 | 3 | 4 | 5 | 6 | 7 |

Strongly
Disagree
　　　　　　　　　　　　　　　　　　　　　　　　　　　　Strongly
　　　　　　　　　　　　　　　　　　　　　　　　　　　　Agree

14. I could easily distinguish among the auditory warnings (i.e., as being an Lane Drift, Forward Collision or Lane Change /Merge Warning).

| 1 | 2 | 3 | 4 | 5 | 6 | 7 |

Strongly
Disagree
　　　　　　　　　　　　　　　　　　　　　　　　　　　　Strongly
　　　　　　　　　　　　　　　　　　　　　　　　　　　　Agree

15. The auditory warnings' tones got my attention.

1   2   3   4   5   6   7

Strongly   Strongly
Disagree   Agree

16. The auditory warnings' tones were not annoying.

1   2   3   4   5   6   7

Strongly   Strongly
Disagree   Agree

17. The yellow lights mounted near the exterior mirrors got my attention.

1   2   3   4   5   6   7

Strongly   Strongly
Disagree   Agree

18. The yellow lights mounted near the exterior mirrors were not annoying.

1   2   3   4   5   6   7

Strongly   Strongly
Disagree   Agree

19. Did the integrated system perform as you expected it to?

Yes_____   No_____

If no, please explain

20. The number of false warnings affected my ability to correctly understand and become familiar with the system

1   2   3   4   5   6   7

Strongly   Strongly
Disagree   Agree

21. The number of false warnings caused me to begin to ignore the integrated system's warnings.

| 1 | 2 | 3 | 4 | 5 | 6 | 7 |

Strongly Disagree　　　　　　　　　　　　　　　　　　　　　　Strongly Agree

22. The integrated system gave me warnings when I did not need them.

| 1 | 2 | 3 | 4 | 5 | 6 | 7 |

Strongly Disagree　　　　　　　　　　　　　　　　　　　　　　Strongly Agree

23. The false warnings were not annoying.

| 1 | 2 | 3 | 4 | 5 | 6 | 7 |

Strongly Disagree　　　　　　　　　　　　　　　　　　　　　　Strongly Agree

24. The integrated system gave me left/right hazard warnings when I did not need them.

| 1 | 2 | 3 | 4 | 5 | 6 | 7 |

Strongly Disagree　　　　　　　　　　　　　　　　　　　　　　Strongly Agree

25. The integrated system gave me left/right drift warnings when I did not need them.

| 1 | 2 | 3 | 4 | 5 | 6 | 7 |

Strongly Disagree　　　　　　　　　　　　　　　　　　　　　　Strongly Agree

26. The integrated system gave me hazard ahead warnings when I did not need them.

1	2	3	4	5	6	7
Strongly Disagree						Strongly Agree

27. How did the false warnings affect your perception of the integrated system?

_____

Overall Acceptance of the Integrated System

28. Please indicate your overall acceptance rating of the integrated system *warnings*
For each choice you will find five possible answers. When a term is completely appropriate, please put a check (√) in the square next to that term. When a term is appropriate to a certain extent, please put a check to the left or right of the middle at the side of the term. When you have no specific opinion, please put a check in the middle.

The integrated system **warnings** were:

useful	☐☐☐☐☐	useless
pleasant	☐☐☐☐☐	unpleasant
bad	☐☐☐☐☐	good
nice	☐☐☐☐☐	annoying
effective	☐☐☐☐☐	superfluous
irritating	☐☐☐☐☐	likeable

assisting	☐☐☐☐	worthless
undesirable	☐☐☐☐	Desirable
raising alertness	☐☐☐☐	sleep-inducing

Displays and Controls

29. The integrated system display was useful.

1   2   3   4   5   6   7

Strongly Disagree   Strongly Agree

30. Did you look at the display less as your experience with the integrated system increased?

Yes_____   No_____

31. The mute button was useful.

1   2   3   4   5   6   7

Strongly Disagree   Strongly Agree

32. The volume adjustment control was useful.

1   2   3   4   5   6   7

Strongly Disagree   Strongly Agree

33. The two lane change/merge warning displays mounted near the exterior mirrors were useful.

| 1 | 2 | 3 | 4 | 5 | 6 | 7 |

Strongly Disagree   Strongly Agree

34. The lane change /merge warnings displays are in a convenient location

| 1 | 2 | 3 | 4 | 5 | 6 | 7 |

Strongly Disagree   Strongly Agree

35. The half circle icons on the center display helped me to understand and to use the integrated system.

| 1 | 2 | 3 | 4 | 5 | 6 | 7 |

Strongly Disagree   Strongly Agree

36. In general, I like the idea of having new technology in my truck.

| 1 | 2 | 3 | 4 | 5 | 6 | 7 |

Strongly Disagree   Strongly Agree

37. Do you prefer to drive a truck equipped with the integrated system over a conventional truck?

Yes_____ No_____

Why?

_____

38. Would you recommend that the company buy trucks equipped with the integrated system?

Yes_____ No_____

Hazard Ahead Warning Acceptance

The Hazard Ahead warning provided an auditory warning accompanied by a brake pulse whenever you were approaching the rear of the vehicle in front of you and there was potential for a collision. When you received this type of warning, the display read "Hazard Ahead."

39. Please indicate your overall acceptance rating of the Hazard Ahead warnings.

For each choice you will find five possible answers. When a term is completely appropriate, please put a check (√) in the square next to that term. When a term is appropriate to a certain extent, please put a check to the left or right of the middle at the side of the term. When you have no specific opinion, please put a check in the middle.

The hazard ahead **warnings** when I was approaching a vehicle ahead were:

useful	☐☐☐☐☐	useless
pleasant	☐☐☐☐☐	unpleasant
bad	☐☐☐☐☐	good
nice	☐☐☐☐☐	annoying
effective	☐☐☐☐☐	superfluous
irritating	☐☐☐☐☐	likeable
assisting	☐☐☐☐☐	worthless
undesirable	☐☐☐☐☐	desirable
raising alertness	☐☐☐☐☐	sleep-inducing

Left/Right Hazard Warning Acceptance

The Left/Right Hazard warning provided an auditory warning whenever your turn signal was on AND you were changing lanes or merging and there was the possibility of a collision with a vehicle in the lane to which you were moving. Or, The Left/Right Hazard warning provided an auditory warning whenever your turn signal was not on and you were drifting out of your lane and there was the possibility of a collision with another vehicle or a solid object (e.g. a guard rail). When you received this type of warning, the display read "Left Hazard" or "Right Hazard" depending on your direction of travel.

40. Please indicate your overall acceptance rating of the Left/Right Hazard warnings.

For each choice you will find five possible answers. When a term is completely appropriate, please put a check (√) in the square next to that term. When a term is appropriate to a certain extent, please put a check to the left or right of the middle at the side of the term. When you have no specific opinion, please put a check in the middle.

The left/right hazard **warnings** were:

useful	☐☐☐☐☐	useless
pleasant	☐☐☐☐☐	unpleasant
bad	☐☐☐☐☐	good
nice	☐☐☐☐☐	annoying
effective	☐☐☐☐☐	superfluous
irritating	☐☐☐☐☐	likeable
assisting	☐☐☐☐☐	worthless
undesirable	☐☐☐☐☐	desirable
raising alertness	☐☐☐☐☐	sleep-inducing

Left/Right Drift Warning Acceptance

If you were drifting out of your lane and there was no danger of you striking a solid object, you received a seat vibration and the display read "Left Drift" or "Right Drift" depending on the direction in which you were drifting.

41. Please indicate your overall acceptance rating of the Left/Right Drift warnings.

For each choice you will find five possible answers. When a term is completely appropriate, please put a check (√) in the square next to that term. When a term is appropriate to a certain extent, please put a check to the left or right of the middle at the side of the term. When you have no specific opinion, please put a check in the middle.

The left/right drift **warnings** were:

useful	☐☐☐☐☐	useless
pleasant	☐☐☐☐☐	unpleasant
bad	☐☐☐☐☐	good
nice	☐☐☐☐☐	annoying
effective	☐☐☐☐☐	superfluous
irritating	☐☐☐☐☐	likeable
assisting	☐☐☐☐☐	worthless
undesirable	☐☐☐☐☐	desirable
raising alertness	☐☐☐☐☐	sleep-inducing

Acceptance of Yellow Lights Mounted Near the Mirrors

When a vehicle was approaching or was in the research vehicle's blind spots, a yellow light in the exterior mirrors was illuminated.

42. Please indicate your overall acceptance rating of the yellow lights in the mirrors.

For each choice you will find five possible answers. When a term is completely appropriate, please put a check (√) in the square next to that term. When a term is appropriate to a certain extent, please put a check to the left or right of the middle at the side of the term. When you have no specific opinion, please put a check in the middle.

The **yellow lights** in the mirrors were:

useful	☐☐☐☐☐	useless
pleasant	☐☐☐☐☐	unpleasant
bad	☐☐☐☐☐	good
nice	☐☐☐☐☐	annoying
effective	☐☐☐☐☐	superfluous
irritating	☐☐☐☐☐	likeable
assisting	☐☐☐☐☐	worthless
undesirable	☐☐☐☐☐	desirable
raising alertness	☐☐☐☐☐	sleep-inducing

Appendix B: Data Processing and Data Mining

Mining the numerical data, coding video data collected during the field test and storing it in a database are essential to the conduct of the independent evaluation. Data mining algorithms were developed to identify and categorize driving conflicts that map to target pre-crash scenarios. A video coding scheme and data logger were created to complement the multimedia data analysis tool (MDAT) in order to quantify information from the video data.

Data Mining

Data mining algorithms were developed to determine the occurrence of driving conflicts and near-crashes in the field test (Lam et al., 2009). Execution of these algorithms created new variables and data structures that were added to the independent evaluation database. The computed variables were developed based on the combination of measured parameters, mathematical computations, and/or equations from previous projects. New data structures, implemented in a Microsoft SQL database, were designed to efficiently store large amounts of driving data.

Data Mining Framework

The data processing framework consists of the following four steps that transform the raw field data into aggregated data of driving conflicts:

Smooth and parse data: This step smoothes the raw data by filling in very short gaps of missing data and filtering noisy data. This step makes the data easier to work with and makes results less erratic. Numerical algorithms for identifying vehicle maneuvers and driving conflicts are then run on the smoothed data to produce these new variables.

- Vehicle maneuvers
  - Vehicle states
  - Vehicle driving states
  - Vehicle maneuvers
  - Vehicle events
  - Driver responses
  - Lane keeping
  - Longitudinal, lateral, and combined motions
- Driving conflicts
  - Closing-in
  - Road and lanes departures
  - Changing lanes or merging

Identify significant events: This step identifies significant events in the conflict driving states. This is followed by numerical analysis of the data to identify false driving conflicts, and/or using the multimedia data analysis tool to verify the occurrence of the conflicts.

Code events: This step codes the significant events in a discrete variable database, after being stored as a continuous stream of sampled data from the previous step. This discrete database consists of conflict, vehicle, and driver files.

Aggregate events: This final step queries the discrete database, using SQL or statistical programs, to aggregate all conflict events. The aggregated driving conflict data are then used by analysts to answer the independent evaluation questions.

Figures 52 and 53 illustrate the process and algorithms used to identify longitudinal and lateral driving conflicts based on raw field test data. The circular blocks represent the input field data. These data are drawn from the radar, in-vehicle, and sensors database. The green blocks denote the algorithms that produce the new variables and their concomitant data summary tables to be added to the independent evaluation database. Finally, the orange blocks refer to the conflict identification algorithms that use the variables created in post processing to determine whether or not a driving conflict has occurred. Rear-end driving conflicts are determined from the 50 percentile near-crash threshold defined in (Najm et al., 2006). The rear-end/LVS scenario was filtered to exclude those based on an LVS event less than three seconds. Moreover, two consecutive longitudinal conflicts were counted as one conflict if they were separated by 2 seconds or less and had the same lead vehicle event. Specific thresholds used to determine conflicts are located in Appendix D.

Table 33 lists the purpose of each of the data mining variables in the block diagrams below. Each variable is created to define a specific aspect of the driving scenario, which is ultimately used to determine the presence of conflicts. Variables are organized by what aspect of the driving scenario they define. Variables defining the host truck motion are created by in-vehicle data and lane-tracking data. Forward target variables are derived primarily from forward-looking radar data. The variable that defines the side target location, adjacent target position, uses the side radar. The road geometry variable is derived from GPS map data. Each conflict variable is calculated using a combination of variables created during data mining.

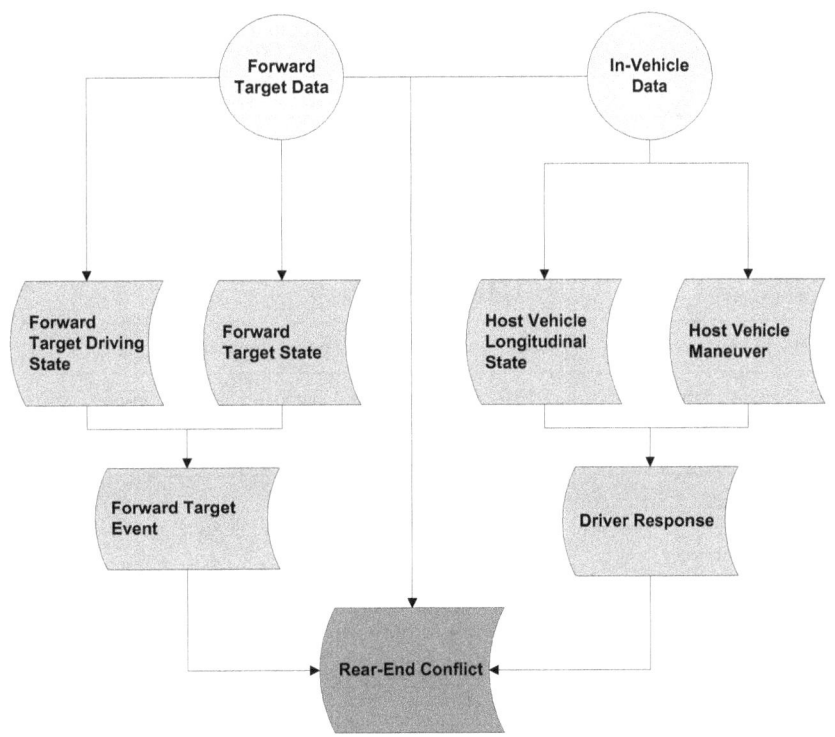

Figure 51. Block diagram of longitudinal driving conflicts

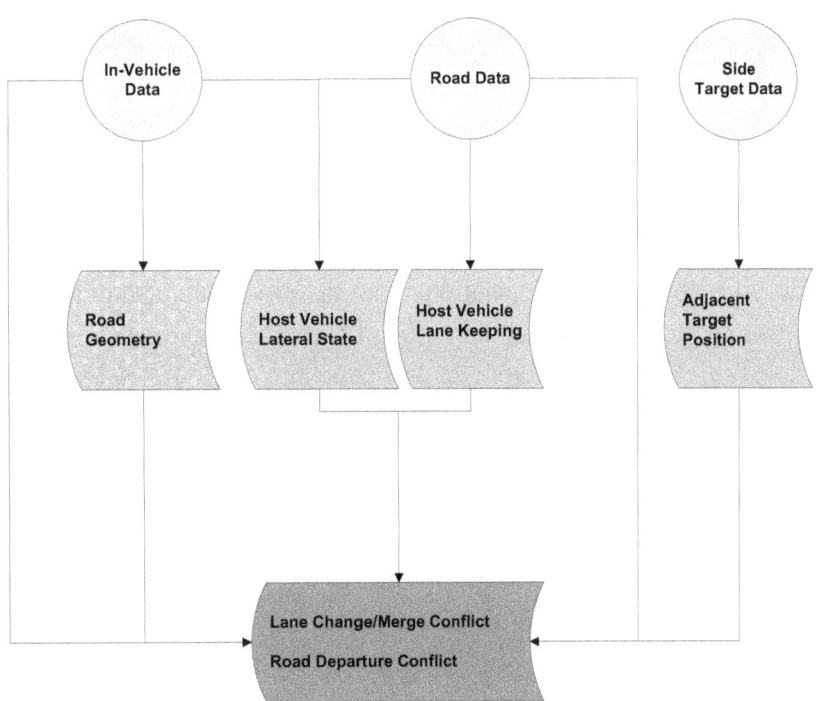

Figure 52. Block diagram of lateral driving conflicts

Table 33. Data mining variables

	Variable	Purpose
Host Truck	Host Vehicle Longitudinal State	Is truck accelerating, decelerating, or constant speed
	Host Vehicle Lane Keeping	Indentifies lane boundary violations
	Host Vehicle Lateral State	Lateral motion of truck
	Driver Response	Driver input to the truck
	Host Vehicle Maneuver	What is truck is doing (lane change, turning, going straight, etc)
Forward Target	In Path Target Count	Determines when the radar detects a new in-path target
	Forward Driving State	Relative speed of the lead vehicle
	Lead Vehicle State	Is lead vehicle accelerating, decelerating, constant speed, or stopped
	Lead Vehicle Category	
	Lead Vehicle Event	Defines events in in which the host vehicle is closing in on the target
Side Target	Adjacent Target Position	Defines relative location of side target
Road	Road Geometry	Determines road type and curvature of the road
Conflicts	Rear End Conflict	Identifies the presence of Rear end conflicts
	Lane Change/Merge Conflict	Identifies the presence of lane change/merge conflicts
	Road Departure Conflict	Identifies the presence of road departure conflicts

Processed Numerical Data

Processing the data mining algorithms resulted in some loss of numerical field data due to missing or invalid information from some trips and data variables. The reduced data set is illustrated in Table 34 by providing information about the processed data in terms of the mileage driven and system alerts received by each test subject during the field test. For comparison, the reader is referred to Table 4 which captures the full exposure of the test subjects. A total of 87,730 miles in baseline and 409,656 miles in treatment were processed, which amount to 68 percent and 87 percent of all exposure data in baseline and treatment test conditions, respectively. For each pick-up and delivery driver, over 92 percent and 86 percent of the data was processed in baseline and treatment test conditions, respectively. As for line-haul drivers, over 60 percent of the baseline data was individually processed for 7 drivers. On the other hand, over 80 percent of the treatment data was processed for 9 of the 10 line-haul drivers.

Table 34. Processed numerical field test data and corresponding system alerts

Driver No.	Baseline					Treatment				
	Miles	FCW	LCM	LDW-I	LDW-C	Miles	FCW	LCM	LDW-I	LDW-C
1	1,522	158	44	117	174	7,741	899	192	690	945
2	2,179	120	98	146	56	7,442	310	441	606	109
4	1,702	137	30	149	47	8,345	845	246	829	162
5	2,271	98	60	161	108	8,709	452	168	502	295
6	2,017	80	75	125	70	2,375	97	84	218	60
7	1,879	152	23	164	228	9,351	691	111	538	879
8	1,527	90	93	211	46	5,141	233	167	305	142
10	1,457	110	66	214	222	7,984	476	201	1,093	1,139
21	8,311	243	79	868	1,185	14,211	257	112	1,356	1,825
22	11,211	251	148	1,337	2,115	57,753	2,165	607	5,565	6,220
23	11,567	329	356	1,200	848	45,450	936	969	5,318	4,582
24	2,899	417	420	704	494	49,227	3,449	1,727	3,077	2,232
25	6,366	272	212	252	237	14,077	438	334	422	289
26	10,371	214	288	788	139	62,103	1,137	1,503	3,849	690
27	7,934	287	323	1,472	639	49,541	741	1,002	3,846	1,672
28	3,735	102	126	792	567	24,365	620	903	3,186	952
29	4,674	79	41	226	260	21,135	568	214	901	722
30	6,108	130	60	248	519	14,705	259	237	564	556
Total	87,730	3,269	2,542	9,174	7,954	409,656	14,573	9,218	32,865	23,471

Appendix C. Post-Drive Survey Mapping to Acceptance Objectives

			Data source												
			Questionnaire										Debrief	Video Analysis	Numerical	Demographic Questionnaire
Objectives	1. Ease of use	1.1 Usability of the warnings	1	2	3	11	12	30	41	42	43	44				
		1.2 Usability of the DVI														
		1.2.i Usability of warning modalities	17	18	19	20	35	36								
		1.2.ii Usability of controls/display	31	37	0											
		1.3 Understanding of the warnings	16	32	0								X			
		1.4 Demands on driver	9											X		
		1.6 Warning patterns													X	
	2. Perceived usefulness	2.1 Usefulness of warnings	4	5	24	26	27	28					X			
		2.2 Safety														
		2.2.i increase in driving safety due to IVBSS	6	10												
		2.2.ii Increase in awareness of surroundings	7													
		2.3 Tolerance of nuisance warnings	0													
		2.3.i Annoyance with nuisance warnings	25													
		2.3.ii Assessment of impact of nuisance warnings	22	23	29								X			
	3. Ease of learning	3.1 Utility of instruction/training														
		3.1.i Time required to become familiar	9													
		3.1.ii Assessment of ability to use IVBSS correctly	21													
		3.2 Comprehension	15													
	4. Advocacy	4.1 Willingness to use IVBSS	39	40												
		4.2 Resistance to new technology	38													
	5. Driving performance	5.1 Control Input														
		5.1.i Snooze button use (frequency/conditions)	33												X	
		5.1.ii Volume use	34												X	
		2.5.2 Vigilance	13	14										X		
Independent Variables	Demographic/ Driving History	Years with CDL														X
		Driving record														X
		Annual mileage														X	X
		Age															X
		Driver experience with IVBSS														X	X
		Prior experience with advanced safety systems?															X
	IVBSS Experience	LDW Availability														X	
		Road Type														X	
		Intensity of Experience															
		Prob of a conflict														X	
		Prob of an alert														X	
		Prob of a coflict/alert														X	
		Prob of alert/conflict														X	
		Occurrence of near crash	9														
		System integration (warning clusters)														X	
		Crash imminent													X		

Appendix D: Video Analysis

A sample of 14,405 heavy-truck videos was selected for the analysis. Each video is associated with a system alert with a duration of 15 seconds, 10 seconds prior to and 5 seconds after the onset of the alert. This time frame encompasses time leading up to the alert to assess the driving scenario and time after the alert to gauge the driver's reaction to the event. The SQL random function was used to select a random sample of alerts from each alert type for each driver. Alerts were also selected proportionally from the baseline and treatment periods. Only alerts with all five videos available were included in the sample.

Table 35 lists the number of alert videos identified by driver and alert type that were analyzed in baseline and treatment test conditions. A total of 6,314 alert videos were analyzed from pick-up and delivery drivers, 30 percent in baseline and 70 percent in treatment. The number of alert videos analyzed from line-haul drivers totaled 8,091 with 27 percent in baseline and 73 percent in treatment. In total, the independent evaluation reviewed 3,861 or 20.4 percent of all FCW alerts, 2,713 or 21.2 percent of all LCM alerts, 4,180 or 9.2 percent of all LDW-I alerts, and 3,651 or 10.6 percent of all LDW-C alerts. Overall, about 13 percent of all 111,860 system alerts issued in the field test in baseline and treatment test conditions were examined.

Table 35. Breakdown of analyzed alert videos

Driver No.	Baseline				Treatment			
	FCW	LCM	LDW-I	LDW-C	FCW	LCM	LDW-I	LDW-C
1	76	22	65	78	215	87	188	205
2	66	58	72	40	149	158	176	73
4	71	22	73	36	208	82	187	96
5	62	40	77	63	173	83	165	124
6	51	48	67	48	76	72	130	62
7	73	17	75	87	197	46	170	196
8	55	55	84	35	129	95	128	81
10	70	44	90	95	177	78	198	195
21	54	32	63	64	118	66	164	171
22	55	47	65	66	174	138	181	182
23	56	57	64	63	159	157	181	181
24	60	59	63	61	179	168	174	168
25	56	53	56	55	137	117	128	111
26	52	54	62	46	163	161	178	143
27	55	55	65	61	153	154	178	165
28	43	40	64	61	149	138	174	158
29	38	23	54	56	155	75	145	138
30	45	30	54	60	112	82	122	127
Total	1,038	756	1,213	1,075	2,823	1,957	2,967	2,576

Video events were coded to collect information that could only be obtained through video analysis of the alert episodes. The specific fields that were coded varied by alert type, based on the type of information that would be necessary to describe the type of driving scenario present for a specific type of alert. The variables are defined so that they can be coded with minimum subjectivity to create consistency in coding across alerts and different reviewers. Recorded numerical data was used to supplement the coded visual information for the analysis of alerts. Table 33 lists the variables that were coded for each alert type. See Appendix C for the coding manual that defines and quantifies the values for each of these variables.

Table 36: Variables coded in video analysis by alert type

All Alerts	FCW	CSW	LDW-I /LCM	LDW-I/LDW-C
Distraction	Target Type	Traverse Curve	Target Type	Lane Excursion Scenario
Eyes off Forward Scene	Target Vehicle Body Type	Passed Road Split	Target Location	Lane Marker
Steering Response	Lead Vehicle Maneuver		Moving Target Vehicle Speed	Road Condition
Host Vehicle Maneuver	Lead Vehicle Position			Opposing Traffic
Host Vehicle Position	In Path of Host Vehicle			Time of Collision
Location	Lead Vehicle Maneuver Times			

Appendix E. Video Coding Manual

INTRODUCTION

This section delineates the variables and codes that were derived from visual observation of video episodes captured during alerts issued by the integrated safety system during the field operational test. The duration of each alert episode is 15 seconds – 10 seconds before the alert onset and 5 seconds after the alert. The list of variables was created to collect information that can only be obtained through video analysis of alert episodes. The variables are defined so that they can be coded with minimum subjectivity to create consistency in the coding across alerts and different reviewers. Numerical data from the data acquisition system will supplement the coded visual information for the analysis of alerts.

VARIABLES AND CODES

The following fields are to be entered based on a review of the video data.

I. All Alerts

The following fields are to be recorded for all alert types:

I.1. *Video Available*:
1. Yes
2. No
3. Not clear

I.2. *Crash Imminent*:
1. No
2. Yes
3. Unsure

I.3. *Distraction*:
1. None
2. Checking blind spot or rear view mirrors
3. Looking to the side/outside car
4. Grooming: High involvement
5. Grooming: Low involvement
6. Eating: Highly Involved
7. Eating: Low involvement
8. Drinking: Highly involved
9. Drinking: Low involvement
10. Adjusting controls
11. Adjusting/using aftermarket device
12. Dialing phone

13. Text messaging
14. Talking/listening to phone
15. Reading Cell Phone
16. Talking/listening to Bluetooth headset
17. Searching interior
18. Reaching for object in vehicle
19. Singing/whistling
20. Talking to/looking at passengers
21. Yawning
22. Eyes closed greater than 1 second
23. Smoking/lighting cigarette
24. Reading
25. Other
26. Unknown

I.4. *Eyes-Off-Forward-Scene*:
1. No (On road)
2. Yes (Off road)
3. Unsure

I.5. *Steering Response*:
1. None
2. Steering before alert
3. Steering after alert
4. Unsure

I.6. *Host Vehicle Maneuver*:
1. Going straight
2. Changing lanes
3. Turning
4. Merging
5. Negotiating curve
6. Other
7. Unsure

I.7. *Host Vehicle Position*:
1. Straight road
2. In curve
3. Curve entry
4. Curve exit
5. Other
6. Unsure

I.8. *Location*:
1. Normal Road
2. Ramp

3. Intersection
4. Normal road AND construction zone
5. Ramp AND construction zone
6. Intersection AND construction zone
7. Unsure

II. FCW Alerts

The following fields are to be recorded for FCW alerts only.

II.1. *Target Type*:
1. No target
2. Moving vehicle
3. Stationary vehicle
4. Roadside sign/object
5. Bridge/overhead sign
6. Guardrail/Jersey barrier
7. Embankment (earth or snow)
8. Pole
9. Other
10. Unknown

II.2. *Target Vehicle Body Type*:
1. No lead vehicle
2. Bicycle
3. Motorcycle
4. Compact/sedan/hatchback
5. SUV
6. Van or minivan
7. Light pickup truck
8. Large truck
9. Bus
10. Other
11. Unsure

II.3. *Lead Vehicle Maneuver*:
1. No lead vehicle
2. Going straight
3. Cut in
4. Cut out
5. Cut in and out
6. Turning off
7. Turning across
8. Cut Across
9. Other
10. Unsure

II.4. *Lead Vehicle Position*:
1. No lead vehicle
2. Straight road
3. In curve
4. Curve entry
5. Curve exit
6. Unsure

II.5. *In Path of Host Vehicle*:
1. Yes
2. No
3. Unsure

II.6. *Eyes on Forward Scene at Lead Vehicle Brake Onset Time*:
0. No
1. Yes
2. Unsure

II.7. *Eyes on Forward Scene at Lead Vehicle Cut-In Time*: Code if Lead Vehicle Maneuver is "Cut in,," "cut in and out,," "turning across," or "cut across.."
0. No
1. Yes
2. Unsure

III. CSW Alerts

The following fields are to be recorded for CSW alerts only.

III.1. *Traverse Curve*:
1. No
2. Yes
3. Unsure

III.2. *Passed Road Split*:
1. No
2. Yes
3. Unsure

IV. LCM/LDW-I Alerts

The following fields are to be recorded for LCM and LDW-I alerts only.

IV.1. *Target Type*:
1. No target
2. Moving vehicle

3. Stationary vehicle
4. Roadside sign/object
5. Bridge/overhead sign
6. Guardrail/Jersey barrier
7. Embankment (earth or snow)
8. Pole
9. Other
10. Unknown

IV.2. *Target Location*:
1. No target
2. Adjacent
3. Two or more lanes over
4. Unsure
5. Adjacent target in forward view

IV.3. *Moving Target Vehicle Relative Speed*:
1. Faster
2. Similar
3. Slower
4. No target vehicle
5. Unknown

V. LDW-C/LDW-I Alerts

The following fields are to be recorded for LDW-C and LDW-I alerts only.

V.1. *Lane Excursion Scenario*:
1. No excursion
2. Intentional excursion/lane change
3. Unintentional excursion
4. Unsure

V.2. *Lane Marker*:
1. Double
2. Single-solid
3. Single-dashed
4. None/barely visible
5. Unknown
6. Other

V.3. *Road Condition*:
1. Dry
2. Wet
3. Snow/slush
4. Salt

5. Unknown
6. Other

V.4 *Opposing Traffic* (left drift only)
1. No opposite direction lane
2. Clear opposite direction lane
3. Occupied opposite direction lane

V.5. ***Time-to-Collision***: Number of data samples between the time the host vehicle first comes into contact with the lane boundary and the time that the vehicle overlaps with opposite direction vehicle in adjacent lane (left alerts only).

EXPLANATION AND CODING INSTRUCTIONS OF VARIABLES AND CODES

This section describes the variables and values presented above. It also provides instructions on how to determine the value for each variable.

I. **All Alerts**

I.1. *Video Available*:

This variable indicates whether or not a video data is available for the alert.
1. Yes - All videos are available and clear
2. No – All 5 videos are missing. If some video channels are present and the others are missing, select "no" if the particular video/videos necessary to analyze the alert is missing (for example, forward video on an FCW alert).
3. Not clear - will be noted for episodes where the video is available and the scene is hard to "see" such as too dark at night.

I.2. *Crash Imminent*:

Watch the full length of the video, paying particular attention to the driver's reaction to the alert. Note whether or not the alert helped the driver avoid a collision and if the alert drew the driver's attention to the hazard. Use your judgment to determine if a crash was imminent at the time of the alert. In any instances of uncertainty, select "Unsure." For episodes coded as "Yes" or "Unsure," the results will be reviewed again by others in order to reach consensus on the final code.

I.3. *Distraction Behavior*:

Pay particular attention to the driver's actions, face, and eyes in the 10 seconds leading up to the alert. Note any distractions that occur any time during this time period from the list below. Select up to three distraction behaviors.
1. None – no obvious distractions.
2. Checking blind spot or rear-view mirrors – driver is looking over shoulder, or in mirrors.
3. Looking to the side/outside cab – driver is looking out windows.

4. Grooming: Low involvement – driver is scratching, running fingers through hair, etc.
5. Grooming: Highly involved – driver is applying makeup, using rearview mirror to look at himself, brushing hair, etc.
6. Eating: highly involved – driver is unwrapping food, eating a sandwich, etc.
7. Eating: Low involvement – driver is eating candy, snacks, etc.
8. Drinking: Highly involved – driver is opening drinks, tipping bottle up to drink.
9. Drinking: low involvement – driver is sipping through a straw, or sipping etc.
10. Adjusting Controls – driver reaching towards center console to adjust in-vehicle controls.
11. Adjusting/using aftermarket device – driver is using device such as navigation system or radar detector.
12. Dialing phone – driver is dialing or pressing buttons on his phone.
13. Text messaging – driver is pressing buttons on his phone, but appears longer than dialing, or is not followed by talking.
14. Talking/Listening to phone – phone visible, listening or talking.
15. Reading cell phone – looking at cell phone but not dialing or talking
16. Talking/Listening to Bluetooth headset – earpiece is in, listening or talking.
17. Searching interior – driver is looking around interior of the vehicle, either front or back seat.
18. Reaching for object in vehicle – driver is retrieving object from somewhere in vehicle.
19. Singing/whistling
20. Talking to/looking at passengers – engaged in conversation with other occupants or looking at/distracted by other occupants.
21. Yawning
22. Eyes closed greater than 1 second – driver's eyes are visibly closed for a period of time longer than one second
23. Smoking or lighting cigarette – cigarette is visible, driver is engaging in any smoking related behaviors including opening window, ashing, smoking, opening cigarette box, etc.
24. Reading – reading material in view, eyes focused toward reading material.
25. Other – any visible distraction that does not fit previous categories.
26. Unknown – video not available.

I.4. *Eyes-Off-Forward-Scene*:

Pay attention to the driver's gaze for the 5-second period before the alert. If the driver's eyes are focused anywhere other than the forward view for a period of at least 1.5 continuous seconds, the driver's eyes are considered to be "off the road." Select "unsure" if it is not possible to tell where the drivers gaze is directed.

I.5. *Steering Response*:

Using the forward view camera and the cabin camera, note whether the driver made any significant steering movements (larger than just minor corrections to remain on current track) just before or after the alert. If the steering correction was initiated at the same time as the alert onset, select "Steering before alert."

I.6. *Host Vehicle Maneuver*:

After watching the videos, make note of any intentional maneuver the driver performed immediately before the alert, or was performing during the time the alert was issued based on the driver's actions, and the front and side view videos. If more than one maneuver occurred, select the maneuver that you feel required the most driver attention. Also, more complicated maneuvers take precedence over less complicated ones. For example, if a driver is passing another vehicle while in a curve, select "Passing" rather than "On a curve."

1. Going straight: Driver travels on a straight road and remains in only one lane, without making any maneuvers.
2. Changing lanes: Driver executes a lane change on a multi-lane road. Directional signals may or may not be used.
3. Turning: Driver is turning or bearing off from one road to another.
4. Merging: Driver is merging into moving traffic on another road, or merging when a lane ends on their current road.
5. Negotiating curve: Vehicle is at any part of a sharp curve in the road, including highway exits or on-ramps and winding roads. This is the same as "going straight" but on a curved road.
6. Other: Other maneuver
7. Unsure: Not sure which maneuver the host vehicle is making

I.7. *Host Vehicle Position*:

Note the position of the host vehicle around the time of the alert.

1. Straight road: Vehicle is traveling on a straight road without intersecting roads.
2. In curve: Vehicle is navigating a curve.
3. Curve entry: Vehicle is approaching or just entering a curve.
4. Curve Exit: Vehicle is exiting a curve or has just completed the negotiation of a curve.
5. Unsure: Unsure of host vehicle position.

I.8. *Location*

Note the location of the host vehicle around the time of the alert.
1. Normal Road: vehicle is driving on a normal road (not a ramp) is not in an intersection and is not in a construction zone.
2. Ramp: Vehicle is navigating a highway on ramp or off ramp.
3. Intersection: Vehicle is passing through an intersection, or is approaching an intersection.
4. Normal road and Construction Zone: Vehicle is traveling through a construction zone where construction or multiple lane markings are visible.
5. Ramp and construction zone: Vehicle is traveling on a ramp that is also a construction zone.
6. Intersection and construction zone: Vehicle is traveling though an intersection where construction is also present.
7. Unsure: Unsure of vehicle location.

II. FCW Alerts

II.1. *Target Type*:

Watch the full length of the video paying particular attention to the forward scene and select the target that is most likely to have triggered the alert (most clearly in front of the vehicle). If the observed target is not on the list, select "Other." If it is not obvious what object caused the alert, select "Unknown." If no target is visible, select "None."

II.2. *Target Vehicle Body Type*:

Identify the type of moveable target that triggered the alert. If "Target type" is not "moving vehicle" select "no lead vehicle."

II.3. *Lead Vehicle Maneuver*:

If the "Target type" has been noted as "Moving Vehicle," note any maneuvers the lead vehicle is making at the time of the alert.
1. No lead vehicle: "Target type" is not "Moving Vehicle."
2. Going Straight: Lead vehicle is traveling in its current lane without making any maneuvers.
3. Cut in: Lead vehicle executes a lane change from an adjacent lane into the lane of travel of the host vehicle, or lead vehicle turns onto roadway in front of host vehicle. Lead vehicle may cut in from the other direction (of what is shown below):

4. Cut out: Lead vehicle executes a lane change to adjacent lane so that they are no longer in the same lane of travel of the host vehicle. Lead vehicle may cut out to the other direction (of what is shown below):

5. Cut in and out: Lead vehicle executes a cut out immediately after a cut in; i.e., moves from one adjacent lane to the adjacent lane on the other side of the vehicle. Lead vehicle may execute this to the other direction (of what is shown below):

6. Turning off: Lead vehicle is preparing to turn onto another road (is slowing), or is turning onto another road. Use blinker to help determine if the lead vehicle intends to turn. Lead vehicle may turn into the other direction (of what is shown below):

7. Turning across: Target vehicle is turning onto a perpendicular road from opposite direction of travel, and passes across path of host vehicle.

8. Cut Across: Target vehicle is traveling across (perpendicular to) the host vehicle's lane of travel at an intersection. Lead vehicle may cut across the other direction (of what is shown below):

9. Unsure: Target

II.4. *Lead Vehicle Position*:

From the forward scene video, determine the characteristics of the road where the lead vehicle is at the time of the alert.
1. No lead vehicle: "Target type" is not coded as "moving vehicle"
2. Straight road: Lead vehicle is traveling on a straight road.
3. In curve: Lead vehicle is navigating a curve.
4. Curve Entry: Lead vehicle is just entering a curve.
5. Curve Exit: Lead vehicle is completing the negotiation of the curve.
6. Unsure

II.5. *In Path of Host Vehicle*:

This variable denotes whether the target is or was in the intended path of the equipped host vehicle (in the lane of travel of the host vehicle) around the alert time. If the vehicle is currently in path at the alert time, or if the vehicle cut in or out of the equipped vehicle path, code as "Yes."

II. 6. *Eyes on Forward Scene at Lead Vehicle Brake Onset Time*:

Note whether the driver's attention was on the forward scene at the time, or within two samples of the lead vehicle's brake onset. Leave blank if there is no brake onset time. Enter the appropriate number into the entry field.
0. No: Driver's attention is not on forward scene
1. Yes: Driver's attention is on forward scene
2. Unsure

II.7. *Eyes on Forward Scene at Lead Vehicle Cut-In Time*:

Note whether the driver's attention was on the forward scene at the time, or within two samples of the time the lead vehicle first begins to enter the host vehicle's lane. Code if Lead Vehicle Maneuver is "Cut in," "cut in and out," "turning across" or "cut across," otherwise leave blank. Enter the appropriate number into the entry field.
0. No: Driver's attention is not on forward scene
1. Yes: Driver's attention is on forward scene

2. Unsure

III. CSW Alerts

III.1. *Traverse Curve*:

From the forward scene, determine if the host vehicle traverses the curve (before, during, or after the alert) for which the warning was issued.
1. Yes: Host vehicle was traversing or entering a curve at the time of the CSW alert.
2. No: Host vehicle was not traversing a curve at the time of, or shortly after a CSW alert.
3. Unsure

III.2. *Passed Road Split*:

If the host vehicle did not traverse a curve, indicate whether or not the vehicle passed a road split where a curve was present (e.g., off ramp)
1. No: split in road
2. Yes: Passed split in road where a curve was present
3. Unsure

IV. LCM/LDW-I Alerts

IV.1. *Target Type*:

Watch the full length of the video paying particular attention to the side scene videos and select the target that most likely to have triggered the alert. If the observed target is not on the list, select "Other." If it is not obvious what object caused the alert, select "Unknown." If no target is visible, select "None."

IV.2. *Target Location*:

The position of side targets with respect to the equipped vehicle at the time of the alert. If no side target or target is unidentifiable, select "No Target." If the equipped vehicle or a POV is changing lanes at the time of the alert, select the lane the vehicle was in before the maneuver.
1. No target
2. Adjacent: Target is adjacent to host vehicle in either a lane of travel, road shoulder, or off the road.
3. Two or more lanes over: There is a full travel lane between the host vehicle and the target.
4. Unknown: Unable to determine the lateral offset of the target.
5. Adjacent target in forward view: Target is in adjacent lane, but is in front of the host vehicle. Rear bumper of vehicle must be visible in the forward camera.

V1.3. *Moving Target Vehicle Relative Speed*:

Note the speed of the lateral moving target relative to the host vehicle at the time of the alert. Use the side and front cameras to determine the relative speed over the length of the video.
1. The target vehicle is traveling faster than the host vehicle.

2. The target is traveling approximately the same speed as the host vehicle.
3. The host vehicle is passing the target vehicle.
4. No target vehicle
5. The relative speed is not determined, or inconsistent over the course of the video.

V. LDW-C/LDW-I Alerts

V.1. *Lane Excursion Scenario*:

By watching the forward and side view videos, indicate the lane keeping behavior of the host vehicle:
1. No Excursion: Vehicle did not leave lane or drift towards lane edge.
2. Intentional Excursion: Driver intentionally swerved out of or to the side of their lane to avoid another vehicle, pedestrian, bicycle, an object in the roadway or driver changes lanes intentionally without turn signal or driver cuts curve to make wide turn or maneuver.
3. Unintentional Excursion: Vehicle leaves lane or drifts towards lane edge, apparently unintentionally, or showing no signs of an intentional maneuver.
4. Unsure: Unclear whether vehicle departed lane or drifted in lane

V.2. *Lane Marker*:

Using the side and forward view cameras, determine the type of lane markings relevant to the side of the alert. Select unknown if the lane marking is undetermined because of poor video quality.

V.3. *Road Condition*:

Using the front view video, indicate the condition of the road surface when alert is issued.
1. Dry: No visible moisture or residue on road
2. Wet: Visible moisture or standing water
3. Snow or slush: Accumulating snow or slush on roadway
4. Salt: Visible salt residue on roadway, possibly obstructing lane lines
5. Unknown: Not able to determine.

V.4. *Opposing Traffic* (left drift only)

This variable makes note of whether a vehicle is approaching in an adjacent, opposite direction travel lane. **This variable should only be coded for left drift alerts.** A lane is considered to have an adjacent opposite direction travel lane only if it is the leftmost travel lane in that direction, and if the road has no barrier (curb, grassy area, Jersey barrier, etc) between the opposite direction lanes.

1. No opposite direction lane – travel lane is not the leftmost travel lane, or there is a divider (curb, grass, jersey barrier, etc.) between the opposite direction roadways.
2. Clear opposite direction lane – lane directly to the left of the travel lane is an opposite direction lane, lane is unoccupied.

3. Occupied opposite direction lane - lane directly to the left of the travel lane is an opposite direction lane, lane has approaching vehicle at any time 10 seconds before or 5 seconds after the alert onset.

V.5. **Time-to-Collision:**

If "Opposing Traffic" is coded "occupied opposite direction lane," count the number of samples between the time the host vehicle first comes in contact with the lane boundary, and the time that the vehicles first meet, or the time their bumpers would come into contact if they were occupying the same lane. Enter the number of samples into the entry field. If the opposing lane vehicle does not meet the host vehicle before the end of the video, leave this field blank.

Appendix F. Conflict Identification Thresholds

The conditions used to determine conflicts analyzed in this document are presented below by conflict type. More details about the data mining procedures used for the data in these analyses can be found in (Najm et al., 2007).

Rear-End Driving Conflicts
Three types of rear-end conflicts are included in this analysis: lead vehicle stopped, lead vehicle decelerating, and lead vehicle moving at slower constant speed. For a conflict to be present, the following criteria must be satisfied:

- Forward target is present (stopped, decelerating, or constant speed) ;
- Driver response present (braking or steering); or
- Time-to-collision (TTC) and range rate within thresholds shown below for each lead vehicle state and driver response.

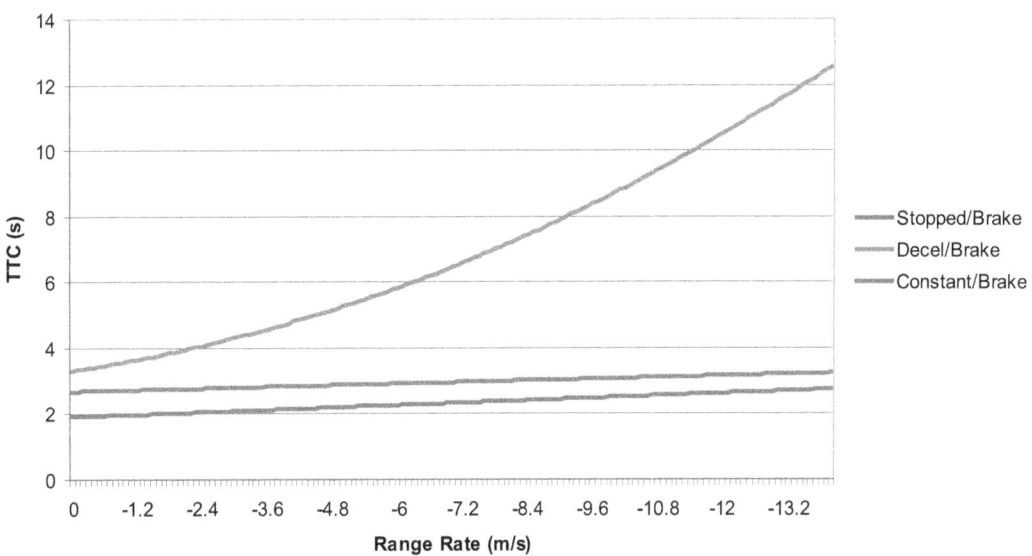

Lane-Change Driving Conflict
The following criteria must be met to determine that a lane-change conflict has occurred:

- Target is present;
- Lane boundary is dashed (not solid);
- Counter-steering response present; and
- Lateral acceleration response (in dir0ection back into lane) greater than 1.2 m/s^2 on a straight road (on curved road, lateral acceleration threshold varies with road geometry).

Road-Departure Driving Conflict

The following criteria must be met for a driving scenario to be considered a road-departure conflict:

- Vehicle crosses solid boundary;
- Counter-steering response present; and
- Lateral acceleration response (in direction back into lane) greater than 1.2 m/s^2 on a straight road (on curved road, lateral acceleration threshold varies with road geometry).

DOT HS 811 464
May 2011

www.ingramcontent.com/pod-product-compliance
Lightning Source LLC
Chambersburg PA
CBHW080256180526
45167CB00006B/2547